在家学做 法式西餐

生活就是一场烹饪

余勇浪 ◎ 著

李英涛 那洪伟 ◎ 译

吉林科学技术出版社

推荐序

法国国际美食协会中国总会长
Mr.Allen Wong

什么是美食，世界性评估机构公告的专业餐厅所供应的是美食，还是网络上的商业机构所推荐的是美食？个人认为都不是，美食是非常独特和个性化的，经其他人舌尖所得出来的意见，不一定能代表个人的喜好和“我的那杯茶”。个人认为，只要个人喜欢，那便是美食，美食是无地域、国界和贵贱之分的！这就是美食的真谛，只要是自己喜欢的便是美食！

认识余勇浪已有几个年头，勇浪是个很年轻的美食“恐怖分子”，他不但全身心投入、深度钻研各种美食素材和烹饪方法，而且极具才华。除了这份对美食的热爱之外，他还经常很热诚地和他身边的人分享他的美食心得和成果。这本《在家学做法式西餐：生活就是一场烹饪》便是一个极好的例子，他把大量的工作经验、研究成果和心得，用由浅入深的方式，让各美食同好把美食带回家和家人分享。在此希望勇浪能继续他的分享，把美食带给每个家庭。

推荐序

食尚小米

世界珠宝小姐中国总冠军　TINA

厨师里最帅的80后型男主厨余勇浪，有着厨师里不多见的帅气，他在《厨王争霸》《中国味道》中展示了自己的精湛厨艺，那些创新的、跳脱出传统的一道道美味就是出自这样一个帅气十足的小伙儿之手。才华和刻苦成就了他，相信这样一个认真的美食工作者会带给您一本全新的美食书籍。

余勇浪是非常努力上进的80后，我和他相识于2013年的电视节目合作上。刚开始并不了解，而经过相识的这么多年，一直见证着他的努力和拼搏，作为朋友我非常开心！继续努力！支持你！

序言

生活就是一场烹饪，走进法式创意西餐

小时候，身边的同学都沉浸在游戏机里，而我，也找到了自己的爱好……就是“吃”。因为父母做饭都很美味，所以我从小学开始就吃到许多美食。然而，对于“吃”这件事情，我不顾父母的反对辍学了，理由是，“吃”，我也会“吃”出一条属于我自己的路。

就这样，我在叛逆的年龄，不对的时间，做出了改变我一生的决定。在2000年15岁初中毕业后，我进入厨房，从最普通的刷盘子、洗菜打杂开始了我的厨艺生涯，曾经有过8年的中餐工作经历，9年的西餐工作经历。

一个骨子里就爱吃的人，在一边工作学习一边吃得多，见得多了以后，味蕾会更加挑剔，做出来的菜也更独一无二、更有自己的见解和风格。

因为我爱吃，也更享受自己亲手料理的美食获得大家喜爱的幸福和满足感。

当然，要把爱好当作职业来做，是需要花费许多时间和精力的，我花了13年的时间在全国各地四处工作学习，从业至今17年，没有国外求学的经历，也没有跟随国内外名厨深造学习过。多年来，

加我新浪微博
@余勇浪-Lance；
朋友们>添加朋友>扫一扫

我四处比赛，也收获了许多奖项。为了追逐梦想，我用自己的方式努力着。

我非常开心能将这本书和热爱美食、热爱生活的朋友分享，怎样在家制作一些美味健康又不复杂的美食并去享受烹饪带来的乐趣。人的一生中的不同时期会有不同的经历，生活终归平淡，却因为有了这些不一样的人生经历而变得丰富多彩、五颜六色。所以，我将这本书命名为《在家学做法式西餐：生活就是一场烹饪》。

我是幸运的，有过近 4 年的担任酒店管理学院酒店管理系特聘讲师的经历。朋友们也会开玩笑说我没有上过大学，却教了那么多大学生。2013年以来，大家都见证了我的成长，我成为了不一样的自己，工作和生活也变得不同。我开始偶尔走出厨房，出现在电视节目上，从事广告、代言、品牌活动等和美食相关的跨界工作。因为跨界也是一件很有趣的事情，可以有更多不一样的人生经历。

很多人都说我，明明可以靠颜值生活，却偏要靠才华。

但是我一直记得我的梦想，我要开一家特别的餐厅，一家自己亲手料理的餐厅。

为了实现它，我必须做特别特别多的努力。

再次感谢每一位出现在我生活中的家人和朋友。

感恩！

生活就是一场极富美感的创意烹饪

法国国际美食会董事会成员
CCTV-2《厨王争霸》两届中国主厨
2015年圣培露全球青年厨师大赛中国前十名
银杏酒店管理学院酒店管理系特聘讲师
美国TEDxSICAU演讲嘉宾
国内外各大知名品牌合作嘉宾及主厨

人们都说厨师赋予了食物第二次生命，余勇浪总是那个将其发挥到极致的人，凭着对中西餐的熟稔，不断创新。走街串巷，寻味不止，只为了打造盘中世界。奖杯荣誉加身，他依然只是那个用心做着美食的人。尊重食材，致敬自然。

10多年来，余勇浪去了很多城市和国家，21岁成为星级酒店的主厨，24岁担任行政总厨，28岁回到成都创业成立工作室至今，是法国国际美食会会员，29岁晋升成为董事会成员。曾受聘成为成都银杏酒店管理学院酒店管理系特聘讲师。

业界年纪尚轻，资历却相当深厚的他，是一位真正的美食创造者，极具人格魅力。正所谓“一位成功者绝不会停留在过去”。每一份菜品都是余勇浪的精心之作，“厨师做菜的机会只有一次”，所以他用绝对严谨的态度来完成每一道菜。

曾经有人问他：如果做出来的菜不符合客人口味怎么办？

“一般不会有这种情况，来吃饭的朋友都特别相信我历练10多年的厨艺，如果客人有疑问，我会以自己的专业去引导他，让他了解所有食材，实在不喜欢那就是客人自己的问题了。”

余大厨还是一位公认的型男主厨

余勇浪的另一个标志性特征便是极为注意自身的形象。平时，余勇浪一年四季总是喜欢西装革履，让自己看起来精神饱满。进

厨房之后，他会换上制服，但是他的制服看起来永远是干净整洁的，自己的仪容看起来也非常得体。

“穿西装的人不怕冷也不怕热，衬衣、马甲、西装、大衣”，无论是出门买菜还是出席活动，他都是一身定制西装，只有做菜时会换厨师服。他喜欢西装的理由很简单，

就是使人看起来精神。

余勇浪颠覆了大家对于厨师这一职业的传统认知与印象，像是在阳光灿烂的午后，他让我们的眼睛和味蕾一同尽情绽放。

当一个男人抛开所有的财富、地位、女人、荣耀……穷得只剩下一套破旧的西服，但他依然自信满满、斗志昂扬的时候，这或许才是真正的男人应该有的样子。西服代表男人不屈的脊梁，余勇浪坚信这一点。

这个有着挺直脊梁骨的男人，比任何人都更能领悟西服对一个男人的意义。他疯狂地爱上了西服这一含蓄收敛、张弛有度、优雅随心的着装。他说，有一种人，他的衣橱里除了西服、衬衣就是领带、马甲、袖扣、口袋巾了，而他就是这种人。他会穿着量身定制的西服出现在超市里、农村菜地、菜市场以及前往泰拳馆的路上，还有各大电视台、广告拍摄现场，法国美食会晚宴……所以，这位国民男神大厨是幸运的，他能在浩瀚寰宇之中找到那种最能表达自我的语言。找到了，他便不会轻易放手。

在《厨王争霸》的舞台上，当所有人被他“最帅主厨”的技艺和外表折服时，他依然想开一家自己料理的餐厅。

所以，余勇浪比别人都努力。

我想要我的生活简单再简单一点儿，把我的生活经营得像是一场极富美感的创意烹饪。

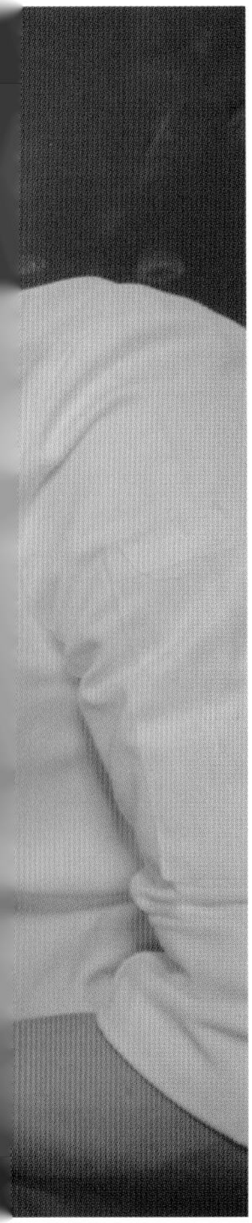
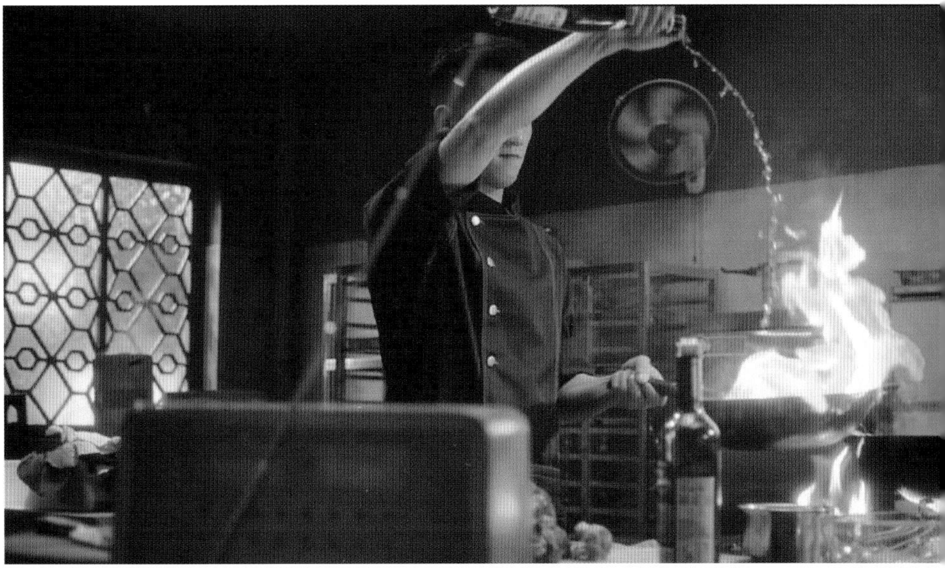
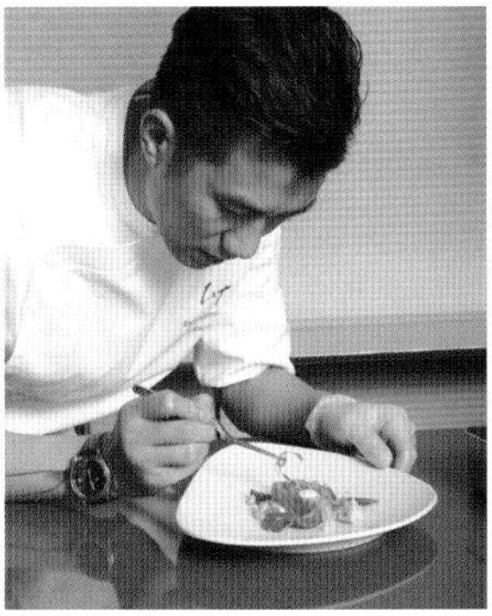
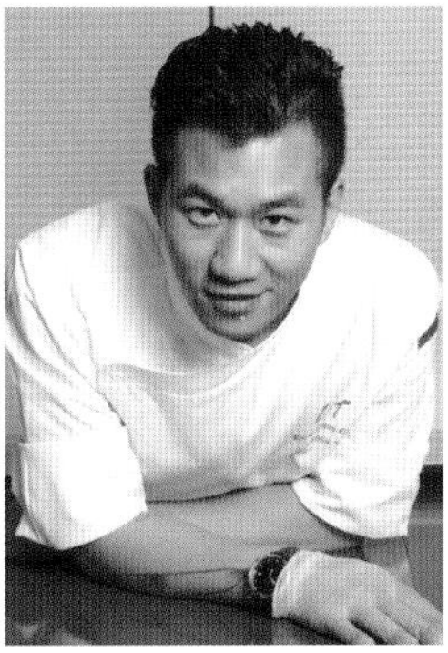
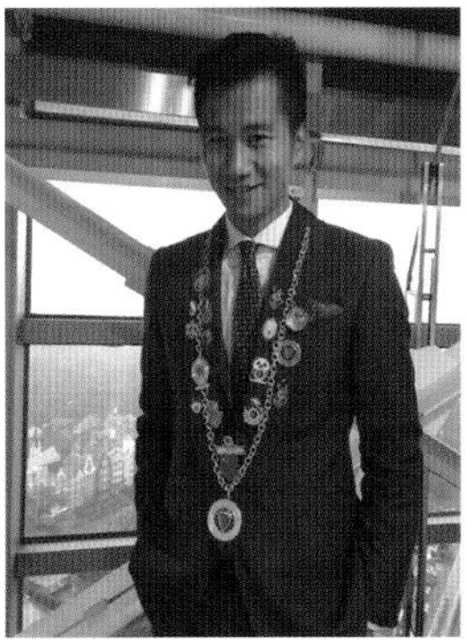
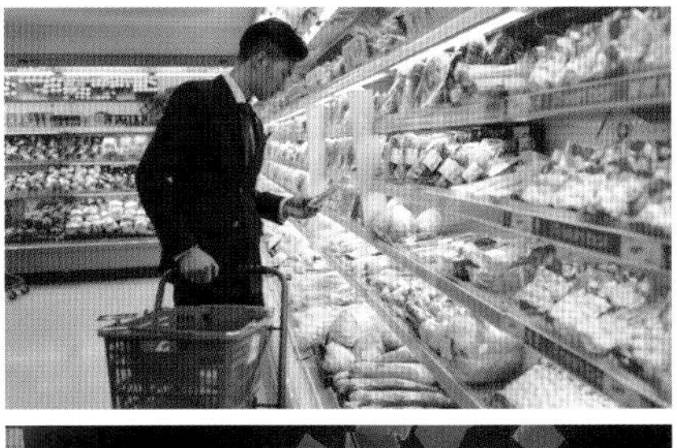
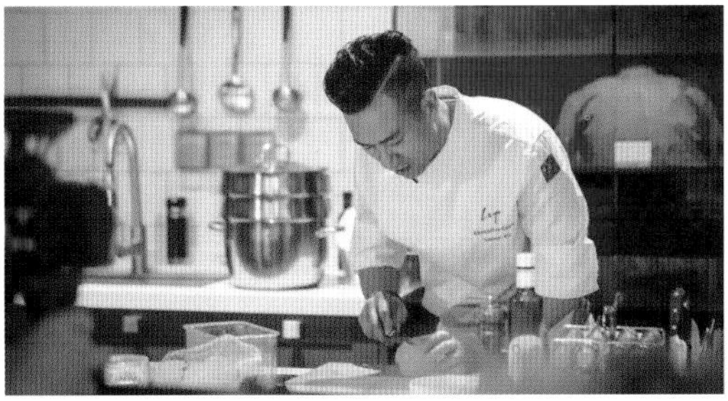
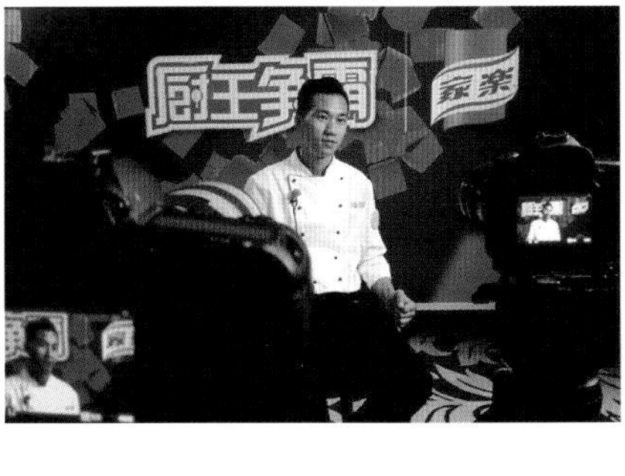
廚王爭霸
家樂
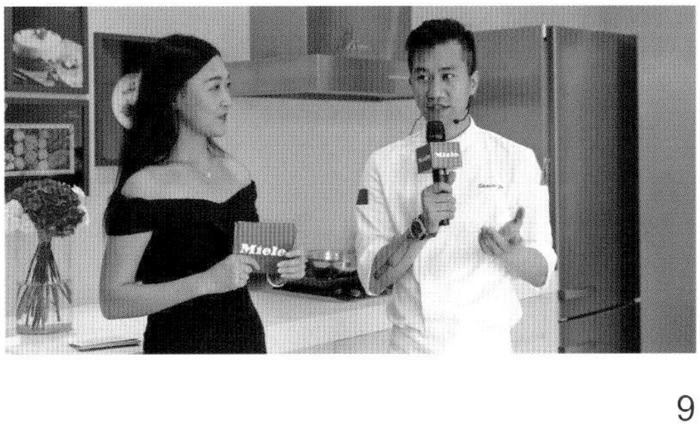
Miele

目录

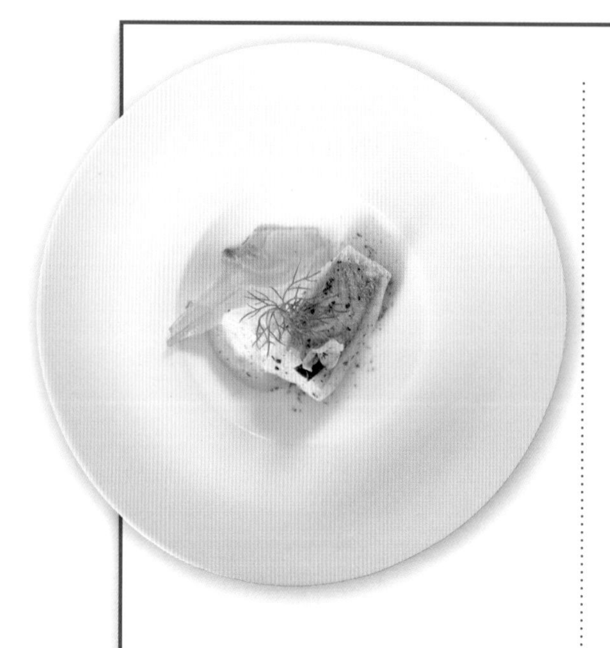

粗盐烤红菜头 / 152
Beet Baked with Coarse sea salt / 152
白酒烩青口贝 / 154
Liquor-braised Mussel / 154
煎牛柳 / 156
Fried Beef Fillet / 156
炒鸡蛋 / 158
Scrambled Eggs / 158
烤澳洲和牛 / 160
Baked Australian Beef / 160
大麦腊肉烩饭 / 162
Rice Braised with Barley and Preserved Meat / 162
竹笋培根烧海参 / 164
Sea Cucumber Braised with Bamboo Shoot and Bacon / 164
油封鸭腿 / 166
Oil-simmered Duck Leg / 166
煎鳕鱼 ♥ 茴香头奶油酱 / 168
Fried Cod Served with Fennel Bulb and Cream Sauce / 168
油封三文鱼芥末酱白酒奶油汁 / 170
Oil-simmered Salmon with Mustard Sauce, white Wine and Cream / 170
奶油水果杯 / 172
Cream and Fruit Cup / 172
焦糖苹果 / 174
Caramelized Apple / 174
焦糖香蕉 / 176
Caramelized Banana / 176
煮鳕鱼 ♥ 南瓜泥焦糖洋葱 / 178
Boiled Cod with Pumpkin Puree and Caramelized Onion / 178
橙子磅蛋糕 / 180
Orange Cake / 180
迷迭香烤苹果 / 182
Rosemary-baked Apple / 182
巧克力草莓 / 184
Chocolate Strawberry / 184
奶油布丁 / 186
Cream Pudding / 186
煎鲈鱼番茄黑橄榄 / 188
Fried Bass with Tomato and Black Olive / 188
炸生蚝 ♥ 迷迭香盐 / 190
Fried Oyster with Rosemary Salt / 190

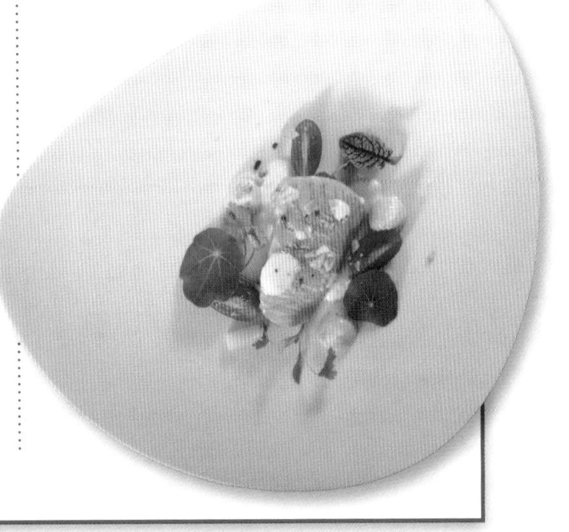

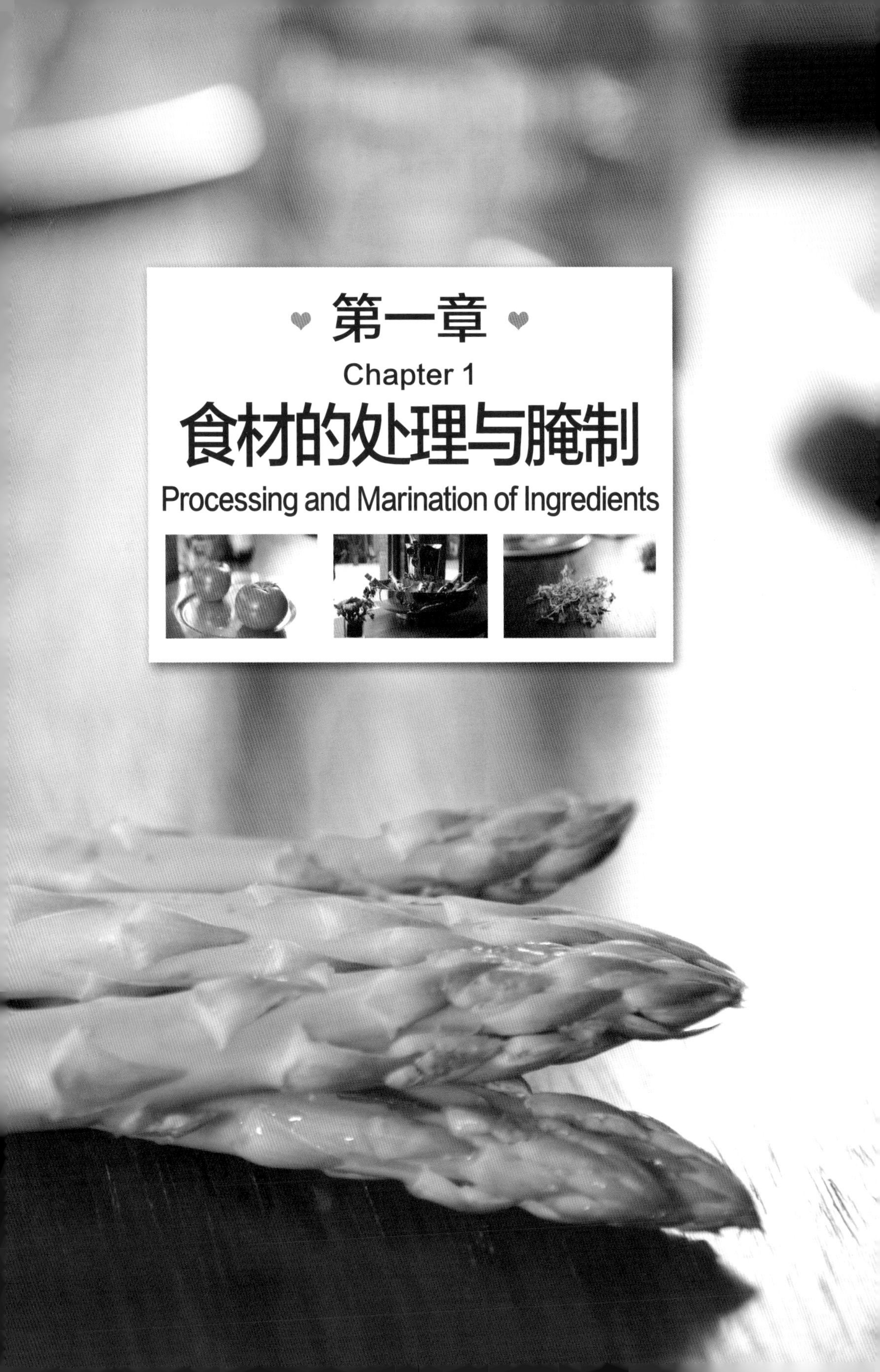

第一章

Chapter 1

食材的处理与腌制

Processing and Marination of Ingredients

常用香草

Common Vanilla

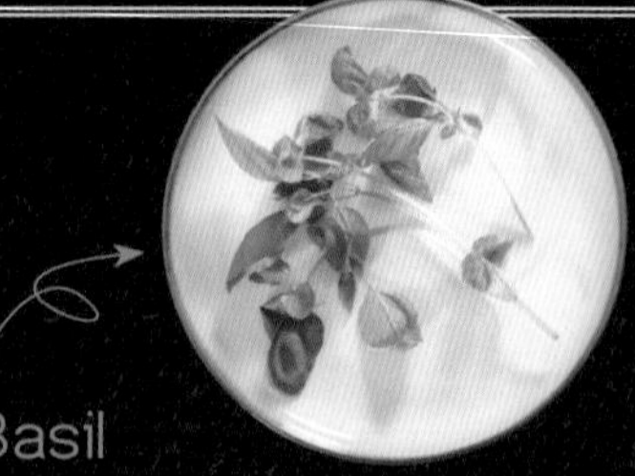

罗勒 Basil

丁香气息，味道辛辣，适合搭配番茄做酱，是意大利面的绝配。

It has the scent of lilac and spicy taste. The sauce made by basil and tomato is a perfect match for pasta.

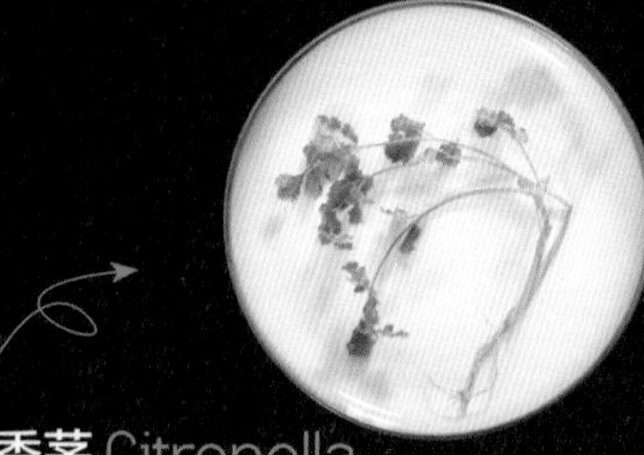

香茅 Citronella

柠檬清香，适用于肉类和海鲜料理、咖喱菜。

It has the scent of lemon, suitable for meat, seafood and curry.

迷迭香 Rosemary

迷迭香——松木香气，清甜中带一点儿苦涩，主要用于烤羊肉、猪肉的腌制。

It has the scent of pine, sweet but a little bitter, mainly used for the pickling of roast mutton and pork.

莳萝 Dill

清凉甘甜，味温和而不刺激，适用于炖类、海鲜佐味。

It is cool, sweet and mild, suitable for stews and seafood.

百里香 Thyme

麝香气息，带点儿清苦，适合与鱼类、贝类食材搭配。

It is a little bitter. With the scent of musk, it is suitable for fish and shellfish.

芝麻菜 Arugula

清嫩口感，带有胡椒与干果的味道，常用来拌沙拉。

It is fresh and tender, has the taste of pepper and dried fruit, often served with salad.

常用水果

Common Fruit

草莓 Strawberry

甜蜜水润，香气扑鼻，适宜搭配味道清淡的食材。

It is sweet and fragrant, best served with light food.

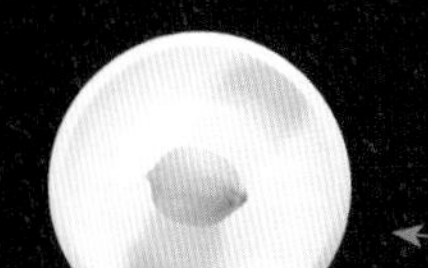

柠檬 Lemon

清新醒胃，平衡油腻，适宜搭配海鲜与肉类。

It is fresh and can balance greasy food, best served with seafood and meat.

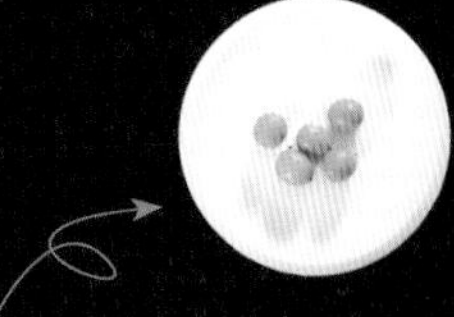

金橘 Kumquat

皮甜肉酸，适宜腌制、做蜜饯，或是炖煮肉类。

It has sweet peel and sour pulp, is suitable for pickling preserved fruit or stewing meat.

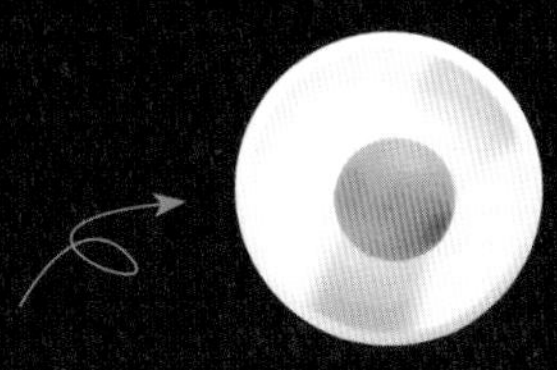

香橙 Orange

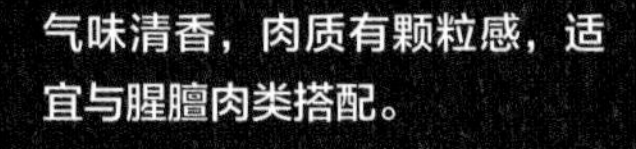

气味清香，肉质有颗粒感，适宜与腥膻肉类搭配。

It is fresh and fragrant, its pulp is full of granules, best served with meat.

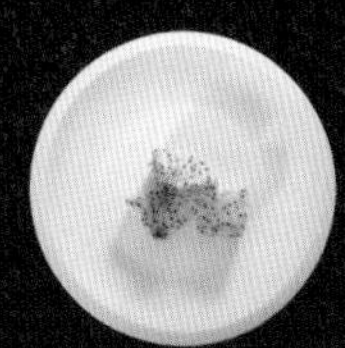

火龙果 Pitaya

滋味清淡，质地紧密，适宜搭配肉类，拌制沙拉。

It tastes fresh, its pulp is tight, best served with meat and salad.

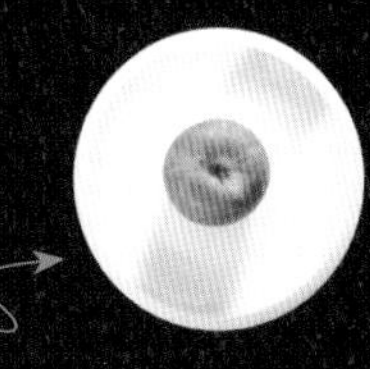

苹果 Apple

酸甜爽脆，解油去腻，适宜搭配肉类，或是熬煮咖喱。

It is sweet and crispy, can balance the greasy food, best served with meat or for cooking curry.

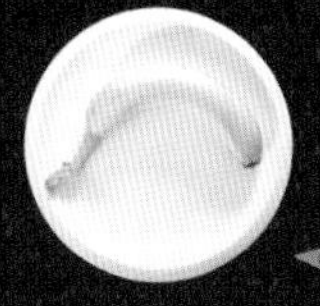

香蕉 Banana

气味浓郁，质地柔软，适宜制作果汁与酱汁。

It is fragrant and soft, suitable for making juice and sauce.

牛油果 Avocado

质地厚实，气味特别，鲜美似奶酪，适宜凉拌，搭配海鲜。

It has thick pulp and special scent, tastes as sweet as cheese and is suitable for making salad or seafood.

猕猴桃 Kiwi

酸甜可口，软硬适中，适宜做沙拉与炒菜。

It is sweet and sour, mild and soft, suitable for salad and fried dish.

羊排的腌制

Marination of Lamb Chop

食材

羊排	200克
迷迭香	1枝
大蒜	2粒
海盐	适量
胡椒	4克
小洋葱	1颗
橄榄油	10毫升
辣椒粉	10克

INGREDIENTS

lamb chop	200 g
rosemary	1 PC
garlic	2 PCs
sea salt	moderate amount
pepper	4 g
small onion	1 PC
olive oil	10 ml
chilli powder	10 g

步骤

① 羊排解冻去掉多余的肥边，切成块。也可以只取肉，不要骨头。

② 腌制：放迷迭香、大蒜、海盐、胡椒、小洋葱、橄榄油、辣椒粉。

③ 搭配后面介绍的酱汁和蔬菜水果一起食用。

PROCEDURES

① Defrost the lamb chop and remove the fat edges, cut the lamb chop into cubes. You can also remove the bones and take the mutton.

② Curing：Put rosemary, garlic, sea salt, pepper, small onions, olive oil, and chilli powder.

③ Best served with the sauce, vegetables and fruits that will be introduced in the following chapters.

三文鱼的腌制
Marination of Salmon

食材

三文鱼	100克
白葡萄酒	10毫升
白胡椒粉	3克
海盐	5克
百里香	2克

INGREDIENTS

salmon	100 g
white wine	10 ml
white pepper	3 g
sea salt	5 g
thyme	2 g

步骤

将切好的三文鱼用适量白葡萄酒、海盐、百里香、白胡椒粉腌制20分钟。

保存方法：加工后的三文鱼可冷藏保存12小时。

注意事项：腌制好的三文鱼应尽快使用。

PROCEDURES

Marinate the chopped salmon with white wine, sea salt, thyme, white pepper for 20 minutes.

Preservation method: Processed salmon can be preserved in refrigerator for 12 hours.

Notes: Marinated salmon should be used as soon as possible.

鲈鱼的腌制
Marination of Bass

食材

鲈鱼…………………… 100 克
柠檬汁…………………… 3 毫升
海盐…………………… 5 克
胡椒粉…………………… 2 克
白葡萄酒…………………… 少许
百里香…………………… 2 克
迷迭香…………………… 2 克

INGREDIENTS

bass …………………… 100 g
lemon juice…………………… 3 ml
sea salt …………………… 5 g
pepper powder…………………… 2 g
white wine …………………… a little
thyme …………………… 2 g
rosemary …………………… 2 g

步骤

1. 鲈鱼去鳞，取出内脏，清洗干净。去掉头、骨，取肉。
2. 腌制：加柠檬汁、海盐、胡椒粉、白葡萄酒、百里香、迷迭香腌制几分钟。
3. 常见烹饪方法：煎、烤、海盐焗。
   可根据后面的介绍，将其与酱汁和蔬菜水果搭配一起食用。

PROCEDURES

1. Remove the scales and internal organs of the bass, wash the bass until clean. Remove the head, bones, and keep the flesh.
2. Curing: Pickle the flesh with lemon juice, sea salt, pepper powder, white wine, thyme and rosemary for a few minutes.
3. Common cooking methods: Fried, grilled, salted.
   It is recommended to match the sauce, the vegetables and fruits that will be introduced in the following chapters.

鳕鱼的腌制
Marination of Cod

食材

鳕鱼块…………………… 100 克
白葡萄酒……………………… 少许
白胡椒粉……………………… 少许

INGREDIENTS

cod…………………… 100 g
white wine ……………… a little
white pepper …………… a little

步骤

将切好的鳕鱼用适量白葡萄酒和白胡椒粉腌制20分钟。
保存方法：加工后的鳕鱼可冷藏保存12小时。
注意事项：腌制好的鳕鱼应尽快食用。

PROCEDURES

Marinate the chopped cod with white wine and white pepper for 20 minutes.
Preservation method: Processed cod can be preserved in refrigerator for 12 hours.
Notes: Marinated cod should be used as soon as possible.

牛肉的腌制

Marination of Beef

食材

牛肉……………………… 200克
迷迭香…………………………… 1枝
小洋葱…………………………… 1个
红酒…………………………… 5毫升
海盐……………………………… 5克
胡椒粉…………………………… 5克
橄榄油……………………… 10毫升

INGREDIENTS

beef ……………………… 200 g
rosemary ………………… 1 PC
small onion……………… 1 PC
red wine ………………… 5 ml
sea salt …………………… 5 g
pepper powder…………… 5 g
olive oil………………… 10 ml

步骤

把全部食材放进容器内，混合均匀，腌制10分钟。煎牛肉时热锅冷油。

PROCEDURES

Put the beef into a container, put the above ingredients in and mix well, fry the beef in 10 minutes. Heat the pan before frying.

英文STEAK一词是牛排的统称，其种类非常多，常见的有以下四种：

★Tenderloin（嫩牛柳，牛里脊），又叫FILLET（菲力），是牛脊上最嫩的肉，几乎不含肥膘，因此很受爱吃瘦肉朋友的青睐。由于肉质嫩，煎到三成熟、五成熟和七成熟皆宜。

★Rib－eye（肉眼牛排），瘦肉和肥肉兼而有之，由于含一定肥膘，这种肉煎烤味道比较香。食用时不要煎得过熟，三成熟最好。

★Sirloin（西冷牛排，牛外脊），含一定肥油，由于是牛外脊，在肉的外延带一圈白色的肉筋，总体口感韧度强、肉质硬、有嚼头，适合年轻人和牙口好的人吃。食用中，切肉时连筋带肉一起切，另外不要煎得过熟。

★T-bone（T骨牛排），呈T字形，是牛背上的脊骨肉。T形两侧一边量多一边量少，量多的是肉眼，量稍少的便是菲力。此种牛排在美式餐厅更常见，由于法餐讲究精致，对于量较大而质较粗糙的T骨牛排较少采用。

牛排几分熟的英文对照：

★Raw，即所谓带血牛肉，是表面稍有一点儿色泽，当中完全是鲜红的生肉状态。

★Rare，即三成熟，切开后仅两个表面是成熟的灰褐色，70%的肉是红色并带有大量血水。

★Medium-rare，五成熟的牛排，50%的肉是红色的，但肉中的血水已较少。

★Medium，七成熟的牛排，切开后，中间断层只有一线红色，肉中血水已干。

★Well done，全熟，熟透的牛排为咖啡色焦黄程度。

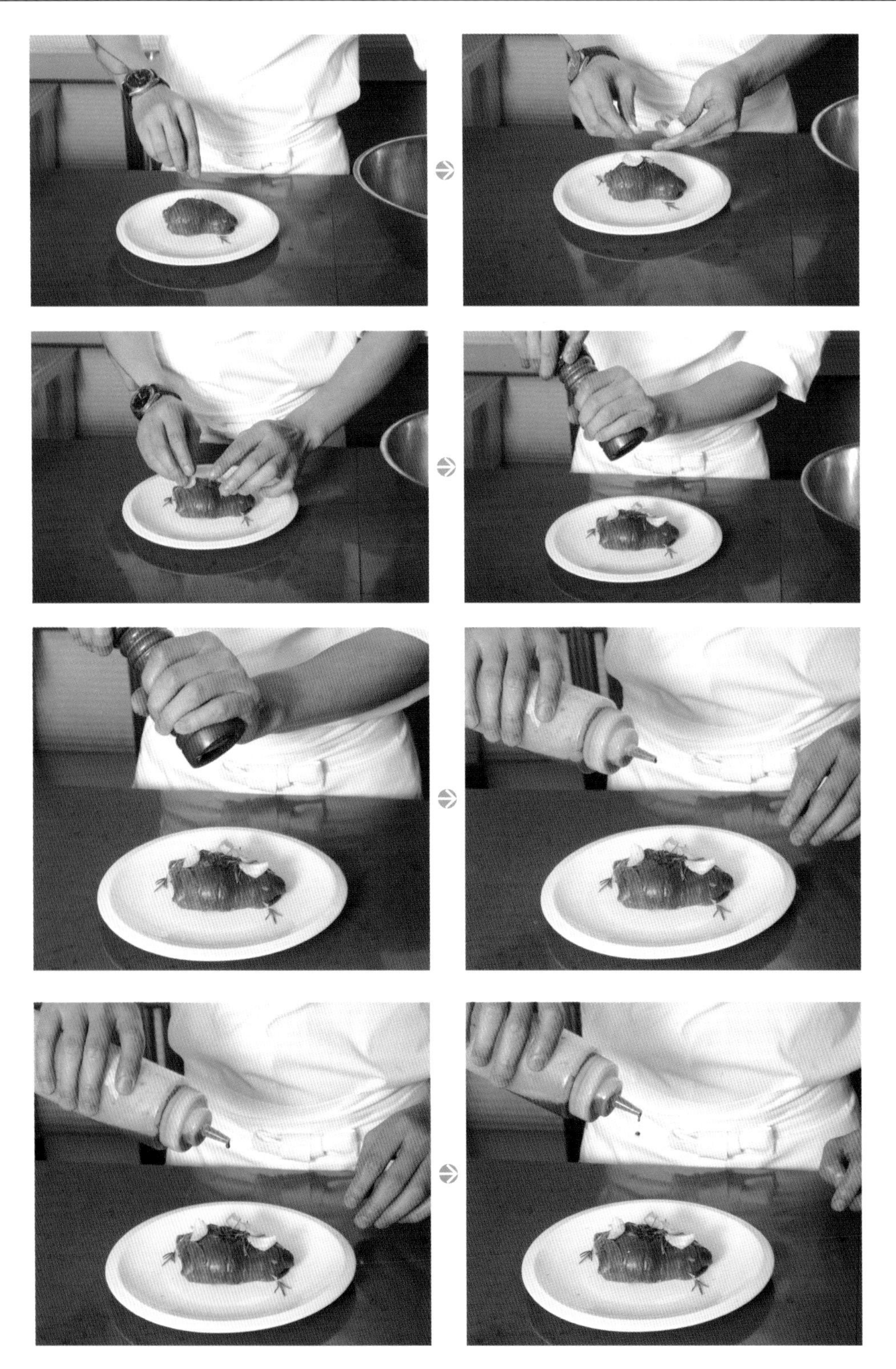

第二章

Chapter 2

汤汁与酱汁

Soup and Sauce

黄色基础汤

Yellow Basic Soup

此汤可用作调汁，烹饪家禽类、海鲜类料理，增加鲜味。

This soup can be used as dressings to cook poultry and seafood to freshen the palate.

猪骨汤 猪骨熬成

Pig Bone Soup

食材	INGREDIENTS
猪骨100克	pig bones 100 g
西芹50克	celery 50 g
胡萝卜50克	carrot 50 g
洋葱20克	onion 20 g
黑胡椒5克	black pepper 5 g
白葡萄酒20毫升	white wine 20 ml
橄榄油30毫升	olive oil 30 ml
百里香5克	thyme 5 g

步骤

1. 猪骨放入水中煮掉血水，冲凉备用。
2. 将橄榄油倒入烧热的汤锅内，加入西芹、胡萝卜、洋葱炒香后，放入白葡萄酒、百里香、黑胡椒，加入煮掉血水的猪骨，加入清水，小火慢慢熬煮。
3. 过滤去渣，即得到猪骨汤。

PROCEDURES

1. Use water to boil pig bones to remove blood and cool off.
2. Pour the olive oil into a hot soup pot, add celery, carrot and onion to fry until fragrant, put in white wine, thyme, black pepper, the pig bones and some water, boil the above ingredients with low flames.
3. Filter the above ingredients to get clear pig soup.

牛肉汤 牛肉熬成

Beef Soup

用作牛肉类食材料理和调牛肉酱汁。

It is used for beef cuisine and beef dressings.

食材	INGREDIENTS
牛肉边角100克	beef edges 100 g
西芹50克	celery 50 g
胡萝卜50克	carrot 50 g
洋葱20克	onion 20 g
黑胡椒5克	black pepper 5 g
白葡萄酒20毫升	white wine 20 ml
橄榄油30毫升	olive oil 30 ml
百里香5克	thyme 5 g

步骤

1. 牛肉边角用烤箱烤焦备用。
2. 将橄榄油倒入烧热的汤锅内，加入西芹、胡萝卜、洋葱炒香后，放入白葡萄酒、百里香、黑胡椒，加入烤箱烤好的边角，放入清水，用小火慢慢熬煮。
3. 过滤去渣，即得到牛肉汤。
   也可在此牛肉汤内加番茄膏，风味会不一样。

PROCEDURES

1. Use oven to bake beef edges to black.
2. Pour olive oil into the hot soup pot, add celery, carrot and onion to fry until fragrant, put in white wine, thyme, black pepper, the beef edges and some water, boil the above ingredients with low flame.
3. Filter the above ingredients to get beef soup.
   Adding some tomato paste to the soup can make the flavor different.

白色基础汤

White Basic Soup

用作海鲜料理的烹饪、调汁。

It is used for the dressing of seafood.

鱼汤 鱼熬成

Fish Soup

食材	INGREDIENTS
鱼骨100克	fish bones 100g
西芹50克	celery 50 g
胡萝卜50克	carrot 50 g
洋葱20克	onion 20 g
黑胡椒5克	black pepper 5 g
白葡萄酒20毫升	white wine 20 ml
橄榄油30毫升	olive oil 30 ml
百里香5克	thyme 5 g

步骤

1. 鱼骨放入水中煮掉血水，冲凉备用。
2. 将橄榄油倒入烧热的汤锅内，加入西芹、胡萝卜、洋葱炒香后，放入白葡萄酒、百里香、黑胡椒，加入煮掉血水的鱼骨，加入清水，用小火慢慢熬煮。
3. 过滤去渣，即得到鱼汤。

PROCEDURES

1. Use water to boil fish bones to remove blood and cool off.
2. Pour olive oil into the hot soup pot, add celery, carrot and onion to fry until fragrant, put in white wine, thyme, black pepper, the fish bones and some water, boil the above ingredients with low flame.
3. Filter the above ingredients to get fish soup.

奶油白汤
Butter and Milk Soup

1

食材	INGREDIENTS
淡奶油100克	unsalted butter 100 g
牛奶50毫升	milk 50 ml
柠檬汁10克	lemon juice 10 g

步骤

牛奶加热，加入柠檬汁，再加淡奶油搅拌。
建议搭配鱼、肉类菜品。

PROCEDURES

Heat the milk, add lemon juice and unsalted butter, and stir well.
Best served with fish and meat.

2

食材	INGREDIENTS
鱼汤500毫升	fish soup 500 ml
淡奶油100毫升	unsalted butter 100 ml
黄油少许	butter a little

步骤

鱼汤和淡奶油混合，收汁变浓稠，加入一些黄油搅拌均匀。
建议搭配海鲜。

PROCEDURES

Mix the fish soup and unsalted butter, concentrate them and add some butter to stir well.
Best served with seafood.

清色基础汤

Clear Basic Soup

清汤用于烹饪蔬菜类，如烩蔬菜等。

It is used for cooking vegetables, stewing vegetables and so on.

蔬菜高汤 蔬菜熬成

Vegetable Soup

食材	INGREDIENTS
食材①	**ingredient 1**
洋葱100克	onion 100 g
番茄200克	tomato 200 g
马铃薯100克	potato 100 g
高丽菜50克	Korean cabbage 50 g
红萝卜50克	red radish 50g
西洋芹60克	celery 60 g
食材②	**ingredient 2**
大蒜1瓣	garlic 1 clove
橄榄油2大匙	olive oil 2 spoons
鸡高汤2毫升	chicken soup-stock 2 ml
豌豆仁50克	pea 50 g
九层塔少许	basil a little
意大利综合香料1小匙	Italian herbs 1 small spoon
海盐适量	sea salt moderate amount
胡椒粉适量	pepper powder moderate amount

步骤

1. 将食材①的所有蔬菜全部切成丁状；食材②的大蒜切末。
2. 将锅预热，放入橄榄油，爆香大蒜末及洋葱丁，放入意大利综合香料、海盐、胡椒粉，略炒一下，再倒进鸡高汤、蔬菜丁、豌豆仁。
3. 用大火煮沸后，将浮沫捞除干净，转小火继续煮约20分钟，盛入汤盘中，上面再放上九层塔装饰即完成。用此道汤品时可撒上少许的芝士粉，会有另一番风味。

PROCEDURES

1. Cut all the vegetables of ingredient 1 into small cubes; cut garlic in ingredient 2 into small pieces.
2. Heat the pan, put in olive oil, fry garlic pieces and onion cubes, add one spoon of Italian herbs, then pour in chicken soup-stock, small vegetable cubes and peas.
3. Boil the above ingredients with high flame and remove the floating foam, then use low flame to boil for 20 minutes, finally put the soup into soup plate and put some basil to decorate. A pinch of cheese powder can be put into the soup to make its flavor different.

基础调料

Basic Seasoning

迷迭香盐

Rosemary Salt

食材	INGREDIENTS
新鲜迷迭香	fresh rosemary
海盐	sea salt

步骤

1. 新鲜迷迭香清洗干净，滤干水分，用烘干机50℃，8小时烘干（也可以采取自然风干，太阳晒干等方式），去掉迷迭香的枝。
2. 迷迭香的叶子加海盐用搅拌机打碎成迷迭香盐。
   搭配红肉类菜品食用，味道更佳。（类似中餐的椒盐做法）

迷迭香盐可以代替普通海盐和酱汁，使菜肴具有更特别的味道。迷迭香盐可用作搭配牛肉类、羊肉类菜肴。

Vanilla salt can replace ordinary sea salt and sauce, adding more flavor to dishes. Rosemary salt can be used in the dishes of beef and mutton.

PROCEDURES

1. Wash fresh rosemary and drain well, put it in a drying machine set for 50 degree for 8 hours (or natural drying, sun drying), and remove its stalks.
2. Use blender to smash the rosemary leaves with sea salt into pieces. Best served with red meat. (It's a little like the making of pepper salt in Chinese food.)

基础油

Basic Oil

迷迭香油

Rosemary Oil

食材 INGREDIENTS

新鲜迷迭香	fresh rosemary
橄榄油	olive oil

步骤

1. 迷迭香洗干净后滤干水分。
2. 橄榄油倒入锅内，烧热至60℃左右，放入洗干净的迷迭香，用微火慢慢熬使其浸出香味。
3. 油和迷迭香一起倒进容器内，一直保存着，慢慢使用。
   搭配红肉类菜品使用味道更佳。

PROCEDURES

1. Wash the rosemary and drain well.
2. Pour olive oil into the pan, heat it to 60 degree, add clean rosemary and simmer until fragrant with low flame.
3. Pour the oil and rosemary into a container. Store it for use.
   Best served with red meat.

迷迭香：其叶带有颇强烈的香味，属于常绿的灌木，夏天会开出蓝色的小花，看起来好像小水滴般，所以 rosmarinus 在拉丁文中是“海中之露”的意思，新鲜的迷迭香也可切碎加入沙拉中进食，在烤羊腿、烤鸡酿馅等菜式中，拌以迷迭香能产生很好的效果。腌肉类原料时也可加入迷迭香以增添香味，是传统羊肉菜肴的上佳搭配。

Rosemary: Its leaves are intensely fragrant. It belongs to evergreen rosebushes. In summer, the rosemary will bloom blue flowers which look like little drops of water, hence rosmarinus means “Dews of sea” in Latin. Fresh rosemary can be cut into small pieces and be put in salad. Better taste can be produced when the rosemary is put into roast lamb leg, roast stuffed chickens and other dishes. Rosemary can also be put into the pickling of meat, and it is an excellent dressing to the traditional mutton dishes.

罗勒油
Basil Oil

食材	INGREDIENTS
新鲜罗勒	fresh basil
橄榄油	olive oil

步骤

1. 罗勒清洗干净后滤干水分。
2. 橄榄油倒入锅内，烧热至60℃左右，放入洗干净的罗勒，用微火慢慢熬使其浸出香味。
3. 橄榄油和罗勒一起倒入容器内，一直保存，慢慢使用。
   搭配海鲜类菜品味道更佳。
   同样的方式可以制作龙虾油、薄荷油、蒜油等，搭配食材之间特性相同的菜肴，增添菜肴的美味。

PROCEDURES

1. Wash the basil and drain well.
2. Pour olive oil into the pan, heat it to 60 degree, add clean basil and simmer until fragrant with low flame.
3. Pour the olive oil and basil into a container. Store it for use.
   Best served with seafood.
   (The same way can be used to make lobster oil, mint oil and garlic oil to make the dishes delicious.)

罗勒在西餐里很常见，和番茄特别搭，潮菜中又名“金不换”。罗勒用于鱼类、野味、家禽、肉类及腌制烧烤食品。我国南北各地，特别是南方及沿海一带均有种植。

Basil is very common in western food. It tastes excellent with tomatoes. It is also known as “Stephania sinica Diels”, used as the dressing of fish, venison, poultry, meat and pickling of baked food. It is planted in all parts of China, especially in the southern and coastal areas.

油醋汁
Vinaigrette

油醋汁可用作蔬菜类沙拉、海鲜类开胃菜。

Vinaigrette can be used in the appetizers like vegetable salad and seafood.

红油醋汁（意大利油醋汁）
Italian Vinaigrette Dressing

食材	INGREDIENTS
芥末5克	mustard 5 g
红葡萄酒10毫升	red wine 10 ml
红酒醋20毫升	red wine vinegar 20 ml
白醋10毫升	white vinegar 10 ml
牛膝草5克	hyssop 5 g
海盐适量	sea salt moderate amount
胡椒10克	pepper 10 g
意大利香醋30克	balsamic vinegar 30 g
青椒30克	green pepper 30 g
红椒30克	red pepper 30 g
洋葱20克	onion 20 g
橄榄油20毫升	olive oil 20 ml

步骤

1. 将红葡萄酒用小火浓缩。
2. 将青椒、红椒、洋葱切成碎粒。
3. 用橄榄油慢慢将芥末顺同一方向调开，加入红酒醋、白醋和浓缩的红葡萄酒、意大利香醋。
4. 加入切好的碎粒和海盐、胡椒、牛膝草即可。
   提示：一般的红油醋汁任意菜式均可以用。

PROCEDURES

1. Concentrate the red wine with low flame.
2. Cut the green pepper, red pepper and onion into pieces.
3. Use olive oil to mix the mustard in one direction, add red wine vinegar, white vinegar, concentrated red wine and balsamic vinegar.
4. Add the above pieces and sea salt, pepper and hyssop.
   Notes: Italian Vinaigrette Dressing can be used in any dish.

白油醋汁（法国油醋汁）
French Vinaigrette Dressing

食材	INGREDIENTS
白葡萄酒20毫升	white wine 20 ml
白酒醋30毫升	white wine vinegar 30 ml
白醋10毫升	white vinegar 10 ml
牛膝草5克	hyssop 5 g
大蒜3克	garlic 3 g
海盐适量	sea salt moderate amount
胡椒5克	pepper 5 g
青椒30克	green pepper 30 g
红椒30克	red pepper 30 g
洋葱20克	onion 20 g
芥末5克	mustard 5 g
橄榄油20毫升	olive oil 20 ml

步骤

1. 将白葡萄酒用小火浓缩。
2. 将青椒、红椒、洋葱、大蒜切成碎粒。
3. 用橄榄油将芥末顺同一方向慢慢调开，加入白酒醋、白醋和浓缩的白葡萄酒。
4. 加入切好的碎粒、海盐、胡椒、牛膝草即可。

   提示：一般油醋汁中醋和油的比例为1：2。

   可搭配任何海鲜和蔬菜类菜肴。

PROCEDURES

1. Concentrate the white wine with low flame.
2. Cut the green pepper, red pepper, onion and garlic into pieces.
3. Use olive oil to mix the mustard in one direction, add white wine vinegar, white vinegar and concentrated white wine.
4. Add the above pieces and sea salt, pepper and hyssop.

   Notes: The proportion of vinegar and oil is 1:2.

   It can be served with any seafood and vegetables.

基础泥
Basic Puree

焦糖辣味花菜泥
Caramelized Spicy Sauce Cauliflower Puree

食材 INGREDIENTS

花菜100克	cauliflower 100 g
白糖50克	sugar 50 g
老抽5毫升	dark soy sauce 5 ml
小米辣1个	capsicum frutescens 1 PC
辣鲜露10毫升	spicy fresh sauce 10 ml

步骤

1. 将白糖放入锅内用小火慢慢炒至呈深咖啡色，加水熬煮成焦糖。
2. 花菜切小朵，放入锅内炒香，小米辣去子，放进锅内，加水，加焦糖熬煮，放老抽、辣鲜露，煮熟花菜。
3. 煮好的花菜和汁倒进搅拌机内打成酱。

建议搭配海鲜类菜品。

PROCEDURES

1. Put sugar into the pan, fry it with low flame to the color of dark coffee, add water, and simmer it to caramel.
2. Cut cauliflower into small pieces, put them into the pan to fry until fragrant. Remove the seeds of capsicum frutescens, put it into the pan, add water, caramel, dark soy sauce and spicy fresh sauce to simmer until the cauliflower is cooked well.
3. Put the cooked cauliflower and juice into blender and smash them into sauce.

Best served with seafood.

花菜泥
Cauliflower Puree

搭配肉类、海鲜类菜肴。

It is used for cooking meat and seafood.

食材 INGREDIENTS

花菜	cauliflower

步骤

花菜切小朵煮熟，打成泥。

PROCEDURES

Cut cauliflowers into pieces , boil them and then smash them into puree.

基础酱汁
Basic Sauce

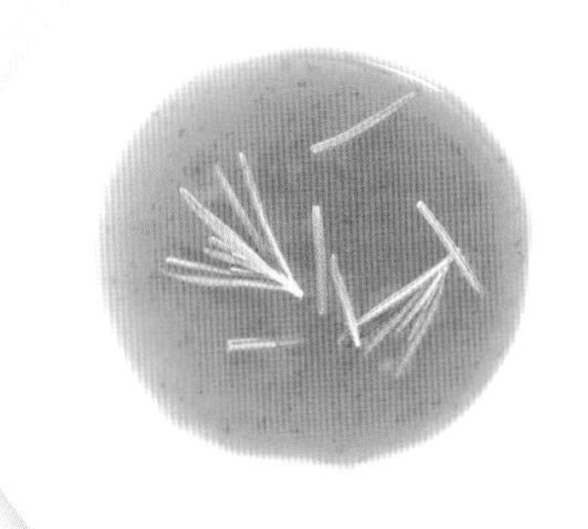

南洋酸辣酱
Nanyang Sour Chilli Sauce

食材	INGREDIENTS
泰式鸡酱150克	Thai chicken sauce 150 g
黑胡椒粉2克	black pepper 2 g
味淋10克	mirin 10 g
葱末8克	minced scallion 8 g
苹果醋5毫升	apple cider 5 ml
蒜末10克	garlic pieces 10 g
柠檬汁20毫升	lemon juice 20 ml
洋葱末15克	onion pieces 15 g
苹果泥20克	apple puree 20 g
米椒末10克	rice chilli pieces 10 g
迷迭香1枝	rosemary 1 PC

步骤

将以上食材（除迷迭香外）混合，放进搅拌机内打碎即可，配以迷迭香。建议搭配海鲜类菜品。冷藏可以长期保存。

PROCEDURES

Mix the above ingredients (except rosemary) and mince them into pieces in a blender and serve with rosemary. Best served with seafood . It can be preserved for a long time via a cold storage.

南洋辣酱
Nanyang Chilli Sauce

食材	INGREDIENTS
泰式鸡酱150克	Thai chicken sauce 150 g
黑胡椒粉2克	black pepper 2 g
味淋10克	mirin 10 g
苹果泥20克	apple puree 20 g

步骤

将以上食材混合，放进搅拌机内打碎即可。冷藏可以长期保存。
建议搭配鸡、鸭、鹅或油炸类菜肴、海鲜类菜品。

PROCEDURES

Mix the above ingredients and smash them into pieces in a blender. It can be preserved for a long time in a fridge.
It is used for cooking chicken, duck, goose, fried food and seafood.

辣卤汁
Chilli Marinades

食材	INGREDIENTS
牛油250克	beef tallow 250 g
菜油100克	vegetable oil 100 g
郫县豆瓣酱150克	broad-bean sauce of Pi county 150 g
豆豉50克	fermented soya beans 50 g
冰糖10克	rock candy 10 g
花椒5克	pepper 5 g
干辣椒30克	dried chilli 30 g
醪糟汁20克	fermented rice wine 20 g
料酒20克	cooking wine 20 g
姜米10克	glutinous rice 10 g
精制海盐100克	refined sea salt 100 g
草果10克	tsao-ko 10 g
桂皮10克	cinnamon 10 g
胡椒粉适量	pepper powder moderate amount
鲜汤1500克	fresh soup 1500 g

步骤

炒锅置旺火上，下菜油烧到六成熟后，下郫县豆瓣酱，姜米、花椒炒香后立即下鲜汤。再放入豆豉、研细的冰糖、牛油、醪糟汁、料酒、海盐、胡椒粉、干辣椒、草果、桂皮。熬开后捞去浮沫即成。

PROCEDURES

Put oil into the heated pan, stir-fry broad-bean sauce of Pi county, glutinous rice and pepper until fragrant, then add fresh soup. Put in fermented soya beans, rock candy powder, beef tallow, fermented rice wine, cooking wine, refined sea salt, pepper powder, dried chilli, tscao-ko and cinnamon, boil the soup and remove the foams.

蜂蜜意大利醋酱
Honey and Italian Vinegar

食材	INGREDIENTS
蜂蜜3匙	honey 3 spoons
意大利醋1瓶	balsamic vinegar 1 bottle
香叶2片	myrcia 2 PCs

步骤

意大利醋倒进锅内，加入香叶、蜂蜜，用小火收浓成酱汁。
建议搭配蔬菜、海鲜类菜品。

PROCEDURES

Pour balsamic vinegar into the pan, add myrcia and honey, concentrate them into sauce with low flame.
Best served with vegetables and seafood.

香橙酱
Orange Sauce

食材	INGREDIENTS
浓缩橙汁100毫升	concentrated orange juice 100 ml
蛋黄酱200克	mayonnaise 200 g
柠檬汁10毫升	lemon juice 10 ml

步骤

将以上食材混合搅拌均匀。
冰箱冷藏可长期保存，建议搭配蔬菜、水果类菜品。

PROCEDURES

Mix the above ingredients and stir well.
The jam can be preserved in refrigerator for a long time. Best served with vegetables and fruits.

红酒酱
Red Wine Sauce

食材	INGREDIENTS
红酒1瓶	red wine 1 bottle
白醋20毫升	white vinegar 20 ml
白糖200克	sugar 200 g
香叶2片	myrcia 2 PCs

步骤

将以上所有食材混合，用小火熬煮至浓稠。
冰箱冷藏可长期保存，建议搭配家禽类、牛肉类菜品。

PROCEDURES

Mix the above ingredients and simmer until dense.
The sauce can be preserved in refrigerator for a long time. Best served with poultry and beef.

海鲜辣酱
Seafood Chilli Sauce

食材	INGREDIENTS
沙司35克	sauce 35 g
辣椒仔5克	chilli pepper 5 g
果糖20克	fructose 20 g
柠檬汁15毫升	lemon juice 15 ml
白兰地2毫升	brandy 2 ml
蒜末6克	garlic pieces 6 g

步骤

将以上食材混合搅拌均匀，
搭配海鲜类菜肴。

PROCEDURES

Mix the above ingredients and stir well.
Best served with seafood.

黑松露牛肉汁
Black Truffle and Beef Sauce

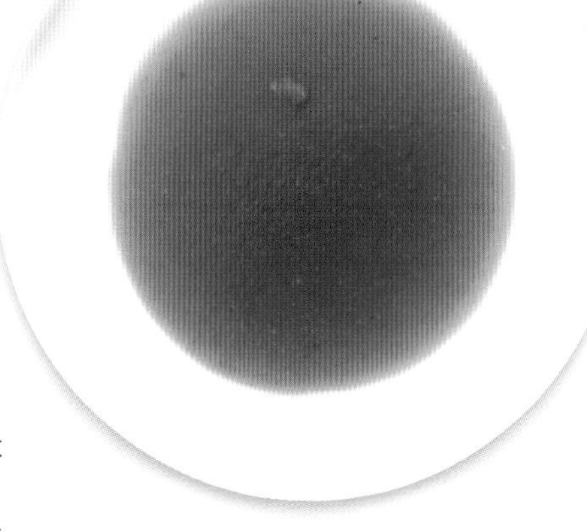

食材	INGREDIENTS
黑松露酱20克	black truffle sauce 20 g
牛肉汤200克	beef broth 200 g
老抽适量	dark soy sauce moderate amount
红酒适量	red wine moderate amount
黄油面粉适量	butter and flour moderate amount

步骤

牛肉汁（参考前面牛肉汤制作介绍）放火上熬煮，加一些红酒烧几分钟后，加入黑松露酱，老抽调色，用黄油面粉收浓。
搭配牛肉。

PROCEDURES

Cook the beef soup (refer to the above mentioned cooking of beef soup), add some red wine , boil for a few minutes, put in black truffle sauce, dark soy sauce, butter and flour.
Best served with beef.

荷兰汁
Hollandaise Sauce

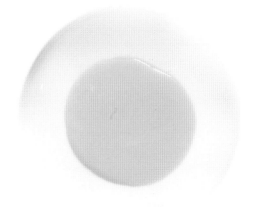

食材	INGREDIENTS
蛋黄2个	egg yolks 2 PCs
清黄油100毫升	butter 100 ml
柠檬汁少许	lemon juice a little

步骤

蛋黄放在碗里，烧一锅热水，把装有蛋黄的碗放在热水锅上，顺时针搅拌，边搅拌边加清黄油，打发后加入柠檬汁，搅拌均匀即成荷兰汁。
搭配鱼肉、芦笋。

PROCEDURES

Put the egg yolks in a bowl, heat a pot of hot water, put the bowl of egg yolks in the pot with hot water, stir clockwise and add butter, then add lemon juice and mix well.
Best served with fish and asparagus.

第三章

Chapter 3

蔬菜与水果的处理

Basic Cooking of Vegetables and Fruits

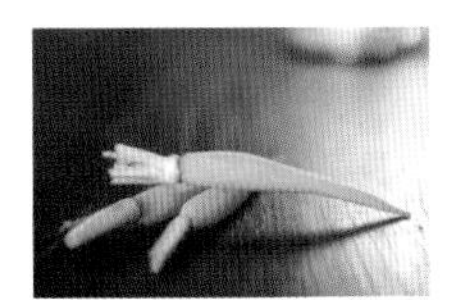

红酒焦糖洋葱
Onion Caramelized with Red Wine

食材

小洋葱……………………… 10 个
红酒……………………… 50 毫升
白糖……………………… 30 克

INGREDIENTS

small onion……………… 10 PCs
red wine ………………… 50 ml
sugar …………………… 30 g

步骤

1. 小洋葱去掉外面一层，切开待用。
2. 锅放在火上，撒上白糖，待白糖焦化时，放入切好的小洋葱。然后倒入红酒，小火熬煮，直到浓稠。
   搭配牛羊肉类的菜品。

PROCEDURES

1. Remove the outer layer of the small onions, cut them into halves for use.
2. Heat the pan, put in sugar, add onions when the sugar caramelizes. Then pour in red wine, and cook until thickened.
   Best served with beef and mutton.

迷迭香烤土豆
Rosemary-baked Potato

食材

土豆	1个
迷迭香	1枝
海盐	10克
黑胡椒	5克
橄榄油	10毫升

INGREDIENTS

potato	1PC
rosemary	1PC
sea salt	10g
black pepper	5g
olive oil	10ml

步骤

① 土豆清洗干净对切成4块，刷一层橄榄油，撒上黑胡椒和海盐，放上迷迭香，用锡纸包裹。

② 烤箱160℃，烤1小时，可根据自己喜好将土豆烤得更干香酥脆。
可搭配牛羊肉菜肴。

PROCEDURES

① Wash the potatoes and cut into 4 pieces, brush a layer of olive oil, sprinkle with pepper and sea salt, put on rosemary and wrap in tin foil.

② Set the oven to 160 degree . Bake for an hour, bake the potatoes to crisp according to one's taste.
Best served with beef and mutton.

土豆泥
Mashed Potato

食材

土豆……………………1个
黄油……………………30克
牛奶……………………50毫升

INGREDIENTS

potato ……………… 1PC
butter ……………… 30g
milk ……………… 50ml

步骤

1. 土豆清洗干净，切小块，放入加有牛奶的水里煮熟。
2. 把煮好的土豆倒进细的油篱里，取一个盆子，用勺子挤压土豆，掉进盆子里会成为非常细滑的土豆泥，然后加入牛奶和黄油搅拌至细滑浓稠。
   可搭配牛羊肉菜品。还可根据自己的口味添加细香葱成香葱土豆泥，加入黑松露酱成松露土豆泥。

PROCEDURES

1. Wash potato and cut it into small cubes and cook them in water with milk.
2. Pour the cooked potatoes into a thin oil hedge, use a spoon to squeeze the potatoes into a basin to get smooth mashed potato. Then add milk and butter, stir to smooth and thick.
   It is recommended to serve beef and mutton. According to personal taste, chives or black truffle can be added into the mashed potatoes.

土豆汤
Potato Soup

食材

土豆……2个
奶油……50毫升
洋葱……10克
培根……20克
海盐……5克
油……20毫升
蔬菜清汤……适量

INGREDIENTS

potato …… 2 PCs
cream …… 50 ml
onion …… 10 g
bacon …… 20 g
sea salt …… 5 g
oil …… 20 ml
vegetable soup … moderate amount

步骤

1. 土豆清洗干净，切小块，洋葱切小块，培根切小片。
2. 锅烧热后，加入油，炒香洋葱和培根，放入土豆、蔬菜清汤（参考前面介绍蔬菜清汤的制作方法），熬煮，加入海盐调味，加入奶油，用搅拌机搅打成土豆浓汤。

PROCEDURES

1. Wash potatoes, cut potatoes and onions into small cubes, and cut bacon into small pieces.
2. After pan is heated, add oil, stir-fry the onion and bacon, add potatoes and vegetable soup (refer to the above mentioned cooking of vegetable soup), boil them, put in dressing and cream, smash them into potato soup in a blender.

梅渍番茄
Prune-soaked Tomato

食材

小番茄……………………500 克
干话梅……………………8 颗
蜂蜜……………………适量

INGREDIENTS

small tomato ……… 500 g
prune ……………… 8 PCs
honey ………… moderate amount

步骤

小番茄在顶部轻轻划一刀，入沸水内煮10秒，倒进冷水里，把皮去掉。开水煮干话梅，放凉开水里加蜂蜜拌匀，把去皮的小番茄放进去。冰箱冷藏保存一天后，取出，会得到非常开胃的梅渍番茄，酸甜可口，非常适合夏季食用或者搭配海鲜类食材。

PROCEDURES

Cut the small tomatoes on the top, boil in water for 10 seconds, then put them in the cold water to peel the tomatoes. Boil prunes, then put them in cold water and mix with honey, put in the small peeled tomatoes, and store them for one day in the refrigerator. The appetizing prune-soaked tomatoes are done. It is very suitable in summer or to be served with seafood.

番茄莎莎
Tomato Baked with Basil Oil

食材

番茄……1个
橄榄油……10毫升
海盐……5克
黑胡椒粉……3克

INGREDIENTS

tomato …… 1 PC
olive oil …… 10 ml
sea salt …… 5 g
black pepper …… 3 g

步骤

将番茄去皮，切细丁，过滤多余番茄汁，加入橄榄油、海盐、黑胡椒粉拌匀。
搭配海鲜、开胃菜作为配菜。

PROCEDURES

Peel the tomato and cut it into small cubes, filter the tomato juice, and add olive oil, sea salt and black pepper to stir well.
Best served with seafood and appetizer as side dishes.

奶油烩蘑菇
Cream-stewed Mushroom

食材

蘑菇	100 克
奶油	20 克
大蒜	2 粒
海盐	5 克
百里香	3 克
白葡萄酒	10 毫升
法香	2 克

INGREDIENTS

mushroom	100 g
cream	20 g
garlic	2 PCs
sea salt	5 g
thyme	3 g
white wine	10 ml
parsley	2 g

步骤

1. 蘑菇清洗干净，大蒜和法香剁碎。
2. 锅烧热，放大蒜炒香，放蘑菇炒到出水，放入奶油炒至溶化，然后加入白葡萄酒煮，放法香、百里香、海盐调味出锅。
   搭配白肉类、家禽野味类菜品。

PROCEDURES

1. Wash the mushrooms, cut garlic and parsley into pieces.
2. Heat the pan, fry the garlic to fragrant. Add the mushrooms and fry them to water. Add the cream to melt. Then add the white wine, parsley, thyme and sea salt and then the dish is done.
   Best served with meat, poultry and venison.

烤小蘑菇
Baked Small Mushroom

食材

小蘑菇…………………… 100克
迷迭香油……………………8毫升
海盐……………………………8克
胡椒粉…………………………5克

INGREDIENTS

small mushroom … 100 g
rosemary oil …………… 8 ml
sea salt ……………………… 8 g
pepper ……………………… 5 g

步骤

小蘑菇去掉蒂部，放进烤盘内，放入其余食材，进160℃烤箱烤至呈金黄色。
搭配牛羊肉类菜品。

PROCEDURES

Remove the end part of small mushrooms, put them into a roast plate, put in the other ingredients, bake them to golden in a 160 degree oven.
Best served with beef and mutton.

酸甜黄瓜
Sour and Sweet Cucumber

食材

黄瓜……………………… 100 克
糖水……………………… 20 毫升
海盐…………………………… 3 克
柠檬汁…………………… 10 毫升

INGREDIENTS

cucumber ………… 100 g
sugar water ………… 20 ml
sea salt ……………………… 3 g
lemon juice …………… 10 ml

步骤

1. 黄瓜刨成片。
2. 把海盐、糖水、柠檬汁混合。
3. 把黄瓜片用调好的汁进行腌渍。

PROCEDURES

1. Cut the cucumber into slices.
2. Mix sea salt, sugar water and lemon juice.
3. Marinate the cucumber slices with the above dressings.

黄瓜卷
Cucumber Roll

食材

黄瓜	1条
海盐	10克
野山椒水	20克
白醋	10毫升
凉开水	100毫升

INGREDIENTS

cucumber	1PC
sea salt	10g
wild mountain pepper water	20g
white vinegar	10ml
cold boiled water	100ml

步骤

1. 黄瓜清洗干净，切小段，用刀片开，不能片断。
2. 凉开水加入海盐、野山椒水、白醋搅匀，放入黄瓜卷浸泡半天。
   按此方法可以泡制各种根茎类蔬菜。（简化版的四川泡菜方法）

PROCEDURES

1. Wash the cucumber, cut it into segments, and blade the segments.
2. Add sea salt, wild mountain pepper water, white vinegar into cold boiled water, and stir well. Put in cucumber rolls and soak for half a day. This method can be used to soak all kinds of stem vegetables.
   (simplified version of Sichuan pickle vegetables)

焦糖苹果
Caramelized Apple

食材

苹果…………………………1个
白糖…………………………少许

INGREDIENTS

apple…………………………1PC
sugar …………………………a little

步骤

1. 苹果去皮、去核，切圆圈，清洗干净。
2. 平底锅烧热，清洗干净的苹果蘸匀白糖，放在平底锅内，小火慢慢煎，直到白糖完全变成焦糖色。
   搭配鹅肝、鸭肝类菜品，有解腻的作用。

PROCEDURES

1. Wash the apple, peel the apple, remove the core of apple, cut it into circles.
2. Cover the apple with sugar, put them into a heated pan, fry with low flame until the sugar turns brown.
   It is recommended to serve goose liver and duck liver. It is used for cleansing the palate.

焦糖香蕉
Caramelized Banana

食材

香蕉	1 根
白糖	少许
柠檬汁	5 毫升

INGREDIENTS

banana	1 PC
sugar	pinch
lemon juice	5 ml

步骤

香蕉去皮对切，在对切面上均匀地撒一层白糖，放进烧热的平底锅内小火慢慢煎，直到白糖变成焦糖色。滴入几滴柠檬汁。
可搭配冰激凌作为甜品。

PROCEDURES

Peel the banana and cut it into halves, cover the sections with sugar, fry in a pan with low flame until the sugar turns brown. Drop some lemon juice.
It is recommended to serve ice creams as a dessert.

波特酒煮金橘
Kumquat Cooked with Port Wine

食材

金橘	100 克
波特酒	20 毫升
蜂蜜	20 克
香草荚	1 枝
水	200 毫升

INGREDIENTS

kumquat	100 g
Port wine	20 ml
honey	20 g
vanilla pod	1 PC
water	200 ml

步骤

金橘清洗干净，放进清水内，加入一些波特酒、蜂蜜和香草荚。小火慢慢熬煮，煮到金橘破皮，关火，一直浸泡着，也可以冷藏保存。

PROCEDURES

Wash the kumquat and put it into clean water, add some Port wine, honey and vanilla pod. Simmer until the kumquat's peel is broken. Turn off the heat, soak the kumquat or preserve in refrigerator.

第四章
Chapter 4
肉类与海鲜的处理
Cooking of Meat and Seafood

煎鳕鱼
Fried Cod

食材

鳕鱼………………………100克
面粉…………………………适量
油…………………………10毫升

INGREDIENTS

cod………………………100 g
flour ……………moderate amount
oil ………………………10 ml

步骤

将鳕鱼切成块，腌好，取出一块腌好的鳕鱼，拍上面粉，放于升温完成的平底锅（涂油）上。双面煎至金黄熟透即可。
注意事项：翻面时不要把鱼弄散了。

PROCEDURES

Cut the cod into pieces to marinate, take out a piece of marinated cod and cover it with flour, and place it on a heated pan (with oil). Fry the cod until both sides are golden yellow.
Notes: Don't scatter the cod when you turn the side.

鳕鱼有带皮和不带皮之分，我们以不带皮为例。

Cook the cod with skin or without skin. We take cooking cod without skin as an example.

煎带子
Fried Scallop

食材

带子……2 个
海盐……3 克
胡椒粉……3 克
柠檬汁……5 毫升
白葡萄酒……4 毫升
橄榄油……少许
黄油……适量
法香碎……少许
土豆慕斯……适量

INGREDIENTS

scallop……2 PCs
sea salt……3 g
pepper……3 g
lemon juice……5 ml
white wine……4 ml
olive oil……a little
butter……moderate amount
parsley……a little
potato mousse moderate amount

步骤

1. 将调味食材（海盐、胡椒粉、柠檬汁、白葡萄酒）混合均匀地抹在带子上，再用厨房纸吸干多余的水分。
2. 平底锅烧热，放橄榄油，放入带子，两面都煎上颜色，加入一些黄油、法香碎，然后出锅。
3. 餐具内放入一些土豆慕斯，依次摆上带子，这样就成为一道开胃海鲜热菜。

PROCEDURES

1. Mix the ingredients (sea salt、pepper、lemon juice、white wine) smear evenly on the scallop then blot up excess water with kitchen paper.
2. Heat up a pan, put olive oil, fry scallop on both sides, add some butter and parsley, then take out.
3. Put some potato mousse in tableware, place the scallop and a hot seafood appetizer is done.

煎三文鱼
Fried Salmon

食材

三文鱼	100克
白葡萄酒	10毫升
白胡椒粉	3克
海盐	5克
百里香	2克
莳萝	3克
橄榄油	10毫升
葱段	4段

INGREDIENTS

salmon	100 g
white wine	10 ml
white pepper	3 g
sea salt	5 g
thyme	2 g
dill	3 g
olive oil	10 ml
green Chinese onion segments	4 PCs

步骤

将三文鱼切成块，腌好，取出一块腌好的三文鱼，放于升温后的平底锅（涂油）上。煎熟至双面有金黄色。

注意事项：翻面时不要把鱼弄散了。

PROCEDURES

Cut the salmon into pieces to marinate, take out a piece of marinated salmon and place it on a heated pan (with oil). Fry the salmon until the two sides turn golden yellow.

Notes: Don't scatter the salmon when you turn the side.

香煎鲈鱼片
Fried Bass Slices

食材

鲈鱼片…………………… 200 克
巴西里末………………………… 少许
马铃薯块………………… 60 克
红萝卜块………………… 60 克
柠檬块…………………… 2 块
面粉……………………… 30 克
奶油…………………………… 2 匙
海盐…………………… 1/2 小匙
胡椒…………………… 1/4 小匙

INGREDIENTS

bass slice ………… 200 g
parsley ……………………… a little
potato cubes …………… 60 g
red radish cubes ……… 60 g
lemon cubes …………… 2 PCs
flour ……………………… 30 g
cream …………………… 2 spoon
sea salt ………… 1/2 spoon
pepper ………… 1/4 spoon

步骤

1. 在鲈鱼片上均匀地撒上海盐和胡椒，再蘸匀面粉，放置约1分钟备用。
2. 将马铃薯块、红萝卜块放入滚水中以大火煮约8分钟，捞出，柠檬挤汁备用。
3. 平底锅烧热，放入奶油烧至溶化，放入鲈鱼片，以中火煎至两面呈金黄色且熟透，取出放入盘中。
4. 将马铃薯块、红萝卜块放入锅中，以中火煎至上色，再加入巴西里末和柠檬汁略炒至香味出来，起锅淋在鲈鱼片上即可。

PROCEDURES

1. Spread sea salt and pepper evenly on the bass slice and then cover with flour for about 1 minute.
2. Put the potato cubes and red radish cubes in boiling water and cook for about 8 minutes, then scoop them out. Squeeze the lemon to get the lemon juice.
3. Heat the pan, add cream and heat it to melt. Put in the bass slice, fry the bass slices with medium flame until two sides are golden yellow. Put the cooked bass slice into a plate.
4. Put the potato cubes and red radish cubes into the pan, fry them with medium flame for some time. Add some parsley and lemon juice and stir-fry until fragrant, turn off the fire and sprinkle them on the bass slice.

腌制鸭腿
Marination of Duck Leg

食材

鸭腿	1个
红酒	10毫升
迷迭香	1枝
海盐	5克
胡萝卜	10克
西芹	10克
洋葱	10克
橙皮	10克

INGREDIENTS

duck leg	1 PC
red wine	10 ml
rosemary	1 PC
sea salt	5 g
carrot	10 g
celery	10 g
onion	10 g
orange peel	10 g

步骤

1. 鸭腿清洗干净，胡萝卜切丝，西芹切细，洋葱、橙皮分别切丝，加入红酒、迷迭香、海盐，混合均匀腌制。
2. 常见烹饪方法：煎、烤、油封。
   可根据自己喜好，选择前面介绍的酱汁和蔬菜水果搭配一起食用。

PROCEDURES

1. Wash the duck leg, cut the carrot, onion, celery and orange peel into shreds, and marinate the duck leg with the above shreds, red wine, rosemary and sea salt.
2. Common cooking methods: Fried, grilled and oil sealed.
   You can choose the sauce, the fruits and vegetables that have been introduced above to serve it.

煎鸭腿
Fried Duck Leg

食材

鸭腿	1 个
面粉	70 克
黄油	70 克
水	1000 毫升
海盐	少许
胡椒粉	少许
橘子	4 只
沙拉油	少许

INGREDIENTS

duck	1 PC
flour	70 g
butter	70 g
water	1000 ml
sea salt	a little
pepper powder	a little
orange	4 PCs
salad oil	a little

步骤

1. 把鸭腿去骨，肉切成块，待用。将橘子去皮，取瓤待用。
2. 在锅中加入鸭骨和水煮开至收浓，鸭汁待用。
3. 在煎锅中放入鸭肉煎上色，待用。
4. 在锅中加入面粉、黄油炒至起沙，再加入煮沸的鸭汁冲匀煮开，并把鸭肉煮熟，然后捞出鸭肉，放在盛器中，并淋入橘子汁，撒上胡椒粉和海盐，即可食用。

PROCEDURES

1. Remove the bones of duck, cut the duck into cubes for use. Peel the orange and take out the flesh for use.
2. Put the duck bone and water in the pot, boil the pot and concentrate the duck soup for use.
3. Fry the duck cubes in a frying pan for some time.
4. Put the flour and butter in the pan and stir-fry. Add the duck soup and boil the duck cubes. Then scoop out the duck cubes, place them in a container, and sprinkle the orange juice.

♥ 第五章 ♥

Chapter 5

法式西餐菜品

French Cuisine

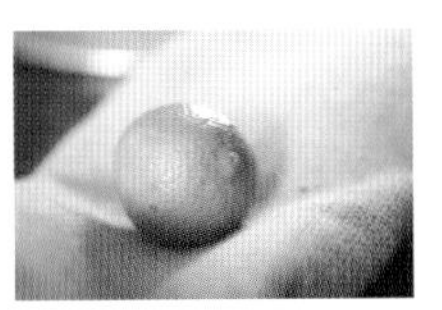
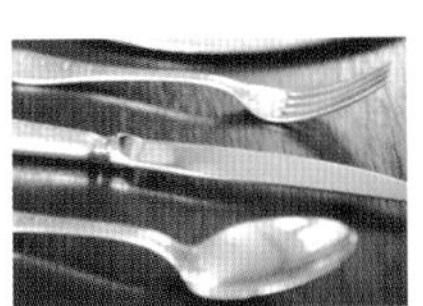

三文鱼牛油果塔

Salmon and Avocado Tower

食材	INGREDIENTS
三文鱼 40 克	salmon 40 g
牛油果 1 / 2 个	avocado half
芝麻菜 2 枝	arugula 2 PCs
胡椒粉 3 克	pepper 3 g
柠檬 1/2 个	lemon half
蜂蜜意大利醋酱	honey Italian vinegar
橄榄油 5 毫升	olive oil 5 ml
海盐 5 克	sea salt 5 g

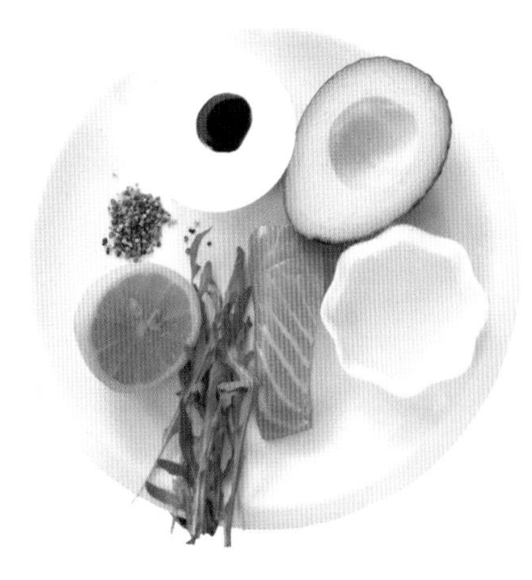

步骤

1. 牛油果对切去核，用勺子挖出牛油果肉，切丁。
2. 三文鱼切丁，与牛油果肉一起放入碗中，放入海盐、橄榄油、胡椒粉调味，用柠檬挤入柠檬汁搅拌均匀。
3. 取一圆形模具，将搅拌好的牛油果、三文鱼放入，压紧，取下模具，周围配蜂蜜意大利醋酱、芝麻菜即可。

PROCEDURES

1. Cut the avocado into halves, remove the core, scoop out the pulp and cut it into small cubes.
2. Dice the salmon, put the salmon and the pulp together in a bowl, add sea salt, olive oil, pepper powder and lemon juice (squeeze the lemon to get lemon juice), stir to mix well.
3. Take a circular mould, put in the mixed avocado and salmon and then press. Take off the mould, add honey Italian vinegar, and match the arugula.

小贴士

1. 三文鱼要自然解冻，鱼刺清除干净。
2. 装盘时可选择自己喜欢的形状摆盘。
3. 蜂蜜意大利醋酱的制作参考前面部分介绍。

Tips

1. Defrost the salmon naturally, and remove all the fish bones.
2. Choose your favorite shape to place the salmon on the plate.
3. Refer to the previous introduction to cook honey Italian vinegar.

三文鱼 · 烤大葱

Salmon and Baked Green Chinese Onion

食材	INGREDIENTS
三文鱼 100 克	salmon 100 g
柠檬 1/2 个	lemon half
糖水煮的金橘	sugar boiled kumquat
大葱 3 段	green Chinese onion 3 segments
西蓝花 2 棵	broccoli 2 PCs
南洋辣酱 20 克	Nanyang chilli sauce 20 g
胡椒粉 2 克	pepper 2 g
海盐 6 克	sea salt 6 g
白葡萄酒 5 毫升	white wine 5 ml
橄榄油适量	olive oil moderate amount

步骤

1. 三文鱼用柠檬汁、白葡萄酒、海盐、胡椒粉腌制，腌制好后放在油锅内，两边煎熟后取出待用。
2. 大葱切段，放入锅中煎至焦糖色，放入油、海盐、胡椒粉调味，盛盘。将三文鱼放在大葱上。
3. 用水煮西蓝花，煮时放入油和海盐调味，煮熟后清炒。
4. 将清炒后的西蓝花和金橘摆在三文鱼周围，配以南洋辣酱即可。

PROCEDURES

1. After pickling salmon with lemon juice, white wine, sea salt and pepper, fry the salmon in a pan until both sides are ripe.
2. Cut the green Chinese onion into segments, fry them in a pan, put in oil, sea salt and pepper, then take them out and put them on a plate. Put the salmon on the green Chinese onion segments.
3. Boil the broccoli with water, put in oil and sea salt, stir-fry the broccoli after it is boiled to ripe.
4. Place the broccoli and kumquat around the salmon, match with Nanyang chilli sauce.

小贴士

1. 南洋辣酱的制作参考前面部分介绍。
2. 大葱必须炒熟，否则会有一点辛辣味。熟的三文鱼容易碎，装盘时要小心。

Tips

1. Refer to the previous introduction to cook Nanyang chilli sauce.
2. The green Chinese onion must be stir-fried until ripe, otherwise it will be a little spicy. Ripe salmon is fragile, take care while placing it on the plate.

煎带子 · 奶油花菜泥

Fried Scallop Served with Caramelized Cauliflower Puree

食材	INGREDIENTS
带子 2 个	scallop 2 PCs
柠檬 1/2 个	lemon half
百里香 1 枝	thyme 1 PC
南洋辣酱 10 克	Nanyang chilli sauce 10 g
花菜适量	cauliflower moderate amount
淡奶油适量	whipping cream moderate amount
海盐适量	sea salt moderate amount
胡椒适量	pepper moderate amount
橄榄油适量	olive oil moderate amount
大蒜少许	garlic a little
白葡萄酒 10 毫升	white wine 10 ml

步骤

1. 带子用厨房纸吸干水分，先撒上海盐、胡椒、白葡萄酒，再用柠檬挤出柠檬汁滴在带子上进行腌制。
2. 腌好后的带子放入锅内，用橄榄油煎，煎至两面金黄时放入大蒜、百里香、淡奶油煎熟。
3. 花菜煮熟打成泥，垫在盘底。
4. 将带子放在花菜泥上，配以南洋辣酱即可。

PROCEDURES

1. After blotting the water of scallop with kitchen paper, sprinkle sea salt, pepper, white wine and drop some lemon juice(squeeze the lemon to get lemon juice) on the scallop, then pickle the scallop.
2. Fry the pickled scallop in a pan until both sides turn golden yellow, put in garlic, thyme and whipping cream, and fry until ripe.
3. Boil the cauliflower to ripe, use a blender to smash it, and place it on the bottom of the plate.
4. Place the scallop on the smashed cauliflowers, match Nanyang chilli sauce.

小贴士

1. 酱汁的制作参考前面部分介绍。

2. 带子要用厨房纸吸干水分，煎的时间不宜过久。

Tips

1. Refer to the above introduction of making sauce.
2. Blot the scallop with kitchen paper, and don't fry the scallop too long.

金枪鱼沙拉

Tuna Salad

食材	INGREDIENTS
金枪鱼 20 克	tuna 20 g
火龙果 30 克	pitaya 30 g
柠檬 1/2 个	lemon half
树莓酱 10 克	raspberry sauce 10 g
西洋菜叶 1 片	lettuce leaf 1 PC
黑胡椒粉 3 克	black pepper 3 g
橙子 1 个	orange 1 PC

步骤

1. 橙子切片，火龙果切丁，金枪鱼切丁。
2. 将柠檬汁、树莓酱和黑胡椒粉搅拌均匀，制成酱汁。
3. 将步骤1中所有食材，淋上酱汁，再配以西洋菜叶即可。

PROCEDURES

1. Slice the orange, dice the pitaya and tuna.
2. Mix them well and put the lettuce leaves at the bottom.
3. Put in tuna and match raspberry sauce.

小贴士

金枪鱼自然解冻，一定要加入柠檬汁杀菌并增添风味。
注意：生食的食材请用专用的菜板和刀具，避免细菌污染。

Tips

Defrost the tuna naturally. Make sure to add lemon juice to kill the bacteria and add flavor.
Notes: In order to avoid bacteria, use special cutting boards and knives to deal with raw ingredients.

步骤

1. 牛舌放入沸水中煮，煮时加入辣椒、花椒、姜、大葱、料酒。煮透放入冰水，去除牛舌上的舌苔。
2. 做一锅四川味的辣卤汁，放入牛舌卤制，卤到牛舌软糯取出，切成块状。
3. 炒少许食材中蔬菜摆盘，放入牛舌。
4. 取少许卤牛舌汁收汁，收汁时加入红酒，将酱汁淋在牛舌上即可。

PROCEDURES

1. Boil the ox tongue, add in chilli, pepper, ginger, green Chinese onion and cooking wine. Place the boiled ox tongue into ice water and remove the fur of the ox tongue.
2. Cook a pot of Sichuan brine, and put in ox tongue to marinate until the tongue is soft, cut it into cubes.
3. Stir-fry some green vegetables, place them on the plate, and put the ox tongue on the top.
4. Concentrate some ox tongue brine, add some red wine during concentration, then sprinkle the concentrated sauce on the ox tongue.

小贴士

1. 辣卤汁的制作参考前面部分介绍。
2. 牛舌一定要卤软烂。

Tips

1. Refer to the previous introduction to cook spicy brine.
2. The ox tongue must be cooked soft and tender.

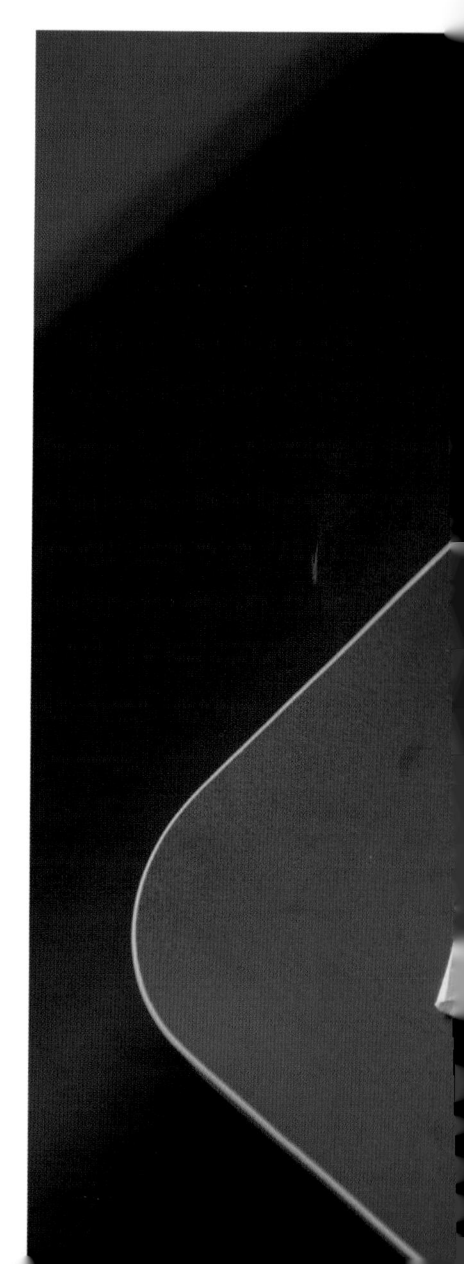

辣卤牛舌

Chilli-marinated Ox Tongue

食材

牛舌适量
芦笋 1 根
胡萝卜球 10 克
茭瓜球 10 克
红酒 10 毫升
辣卤汁 1 锅
辣椒适量
花椒少许
姜少许
大葱少许
料酒适量

INGREDIENTS

ox tongue moderate amount
asparagus 1 PC
carrot ball 10 g
marrow ball 10 g
red wine 10 ml
sichuan brine 1 pot
chilli moderate amount
pepper a little
ginger a little
green Chinese onion a little
cooking wine moderate amount

芦笋 · 荷兰汁

Asparagus Served with Hollandaise Sauce

食材 INGREDIENTS

食材	INGREDIENTS
荷兰汁适量	Hollandaise sauce moderate amount
芦笋若干根	asparagus several
牛奶 50 毫升	milk 50 ml
海盐少许	sea salt a little

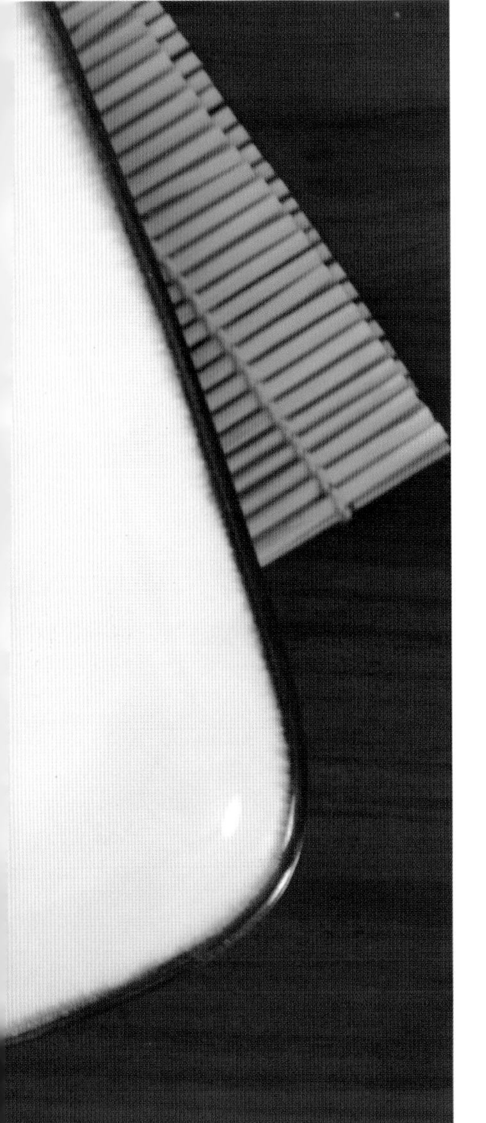

步骤

1. 芦笋去皮，锅中加入牛奶和水。
2. 起火，将芦笋放入锅中煮透，再加入适量的海盐，煮好后放入盘中。
3. 将做好的荷兰汁淋在芦笋上。

PROCEDURES

1. Peel the asparagus, add milk and water into the pan.
2. Heat the pan, put in the asparagus, and cook it to ripe, add some sea salt and place it on the plate.
3. Sprinkle the Hollandaise sauce on the asparagus.

小贴士

1. 荷兰汁的制作参考前面部分介绍。
2. 芦笋煮至断生，保持脆嫩，颜色翠绿。

Tips

1.Refer to the previous introduction to cook Hollandaise sauce.
2.While cooking the asparagus, keep the color green.

步骤

1. 芦笋去皮，放在蔬菜高汤内煮透，取出。
2. 撒上黑胡椒、海盐、迷迭香，放入烤箱中烤制，烤到芦笋皮焦黄时取出。
3. 再次撒入迷迭香及海盐即可。

PROCEDURES

1. Peel the asparagus, cook them in vegetable soup-stock and take them out.
2. Sprinkle black pepper, sea salt and rosemary on the asparagus, then bake it in the oven till it turns yellow.
3. Sprinkle rosemary and sea salt.

小贴士

1. 蔬菜高汤的制作参考前面部分介绍。
2. 芦笋烤的时间不宜太久。

Tips

1. Refer to the previous introduction to cook vegetable soup.
2. Asparagus should not be baked for long.

迷迭香烤芦笋

Asparagus Served with Hollandaise Sauce

食材

迷迭香 1 枝
芦笋 3 根
迷迭香盐适量
黑胡椒 2 克
海盐 2 克
蔬菜高汤适量

INGREDIENTS

rosemary 1 PC
asparagus 3 PCs
rosemary salt moderate amount
black pepper 2 g
sea salt 2 g
vegetable soup-stock moderate amount

温泉蛋 · 煮芦笋

Boiled Egg and Boiled Asparagus

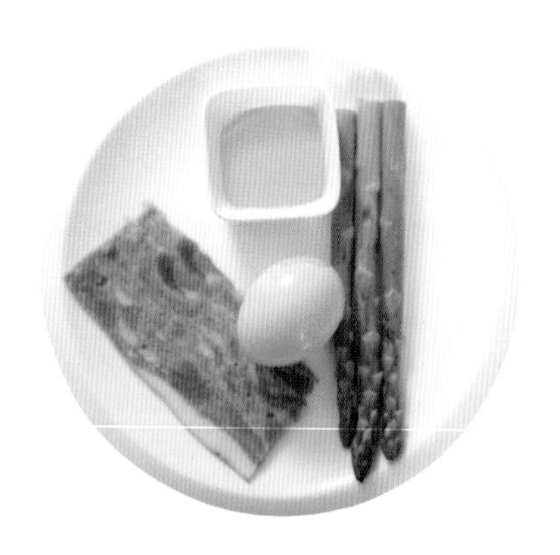

食材	INGREDIENTS
芦笋若干根	asparagus several
鸡蛋 1 个	egg 1 PC
培根 10 克	bacon 10 g
黄油 10 克	butter 10 g
蔬菜高汤适量	vegetable soup-stock moderate amount

步骤

1. 芦笋去皮，加入溶化的黄油，一起放入蔬菜高汤内煮透取出。
2. 将培根切碎，炒好备用。
3. 鸡蛋放在65° 热水中煮45分钟。
4. 将煮好的鸡蛋去皮，放在芦笋上，配以炒好的培根碎。

PROCEDURES

1. Peel the asparagus, add melted butter, cook vegetable soup-stock and take them out.
2. Cut the bacon into pieces and stir-fry them.
3. Cook the egg in 65 degree water for 45 minutes.
4. Peel the cooked egg, put it on the asparagus, and serve cooked bacon pieces.

小贴士

1. 鸡蛋也可以冷水下锅，煮 5 分钟。
2. 蔬菜高汤的制作参考前面部分介绍。

Tips

1. The egg can be cooked in cold water for 5 minutes.
2. Refer to the previous introduction to cook vegetable soup-stock.

脆皮鹌鹑蛋·绿芦笋荷兰汁

Crispy Quail Eggs Served with Green Asparagus and Hollandaise Sauce

食材	INGREDIENTS
绿芦笋 3 根	green asparagus 3 PCs
鹌鹑蛋 3 个	quail eggs 3 PCs
沙拉油 300 毫升	salad oil 300 ml
荷兰汁适量	Hollandaise sauce moderate amount
海盐 10 克	sea salt 10 g
牛奶适量	milk moderate amount

步骤

1. 芦笋削皮，放入水中，加牛奶和海盐煮15秒，煮到颜色更绿后捞出。
2. 鹌鹑蛋煮熟去壳，放入油锅高温炸至金黄色捞出，放入芦笋中间。
3. 将荷兰汁淋在鹌鹑蛋和芦笋上。

PROCEDURES

1. Peel the asparagus, put them into water, put in milk and sea salt cook for 15 seconds. After the color turns greener, take them out.
2. Boil the quail eggs and remove the shells, fry the eggs until golden yellow, and put them between the asparagus.
3. Sprinkle the Hollandaise sauce on the quail eggs and asparagus.

小贴士

1. 荷兰汁的制作参考前面部分介绍。
2. 鹌鹑蛋煮熟后，下高温油锅内炸成脆皮，注意安全。

Tips

1.Refer to the previous introduction to cook Hollandaise sauce.
2.Be careful when frying the boiled quail eggs in a hot pan with oil.

奶油蘑菇汤

Creamed Mushroom Soup

食材	INGREDIENTS
泡水干香菇 1500 克	dry mushroom 1500 g
鲜香菇 480 克	fresh mushroom 480 g
蘑菇 600 克	mushroom 600 g
淡奶 500 毫升	evaporated milk 500 ml
口蘑 480 克	apricot mushroom 480 g
洋葱 1000 克	onion 1000 g
大葱适量	green Chinese onion moderate amount
鸡粉 20 克	chicken powder 20 g
泡香菇水 6000 毫升	soaked water of dry mushrooms 6000 ml
黄油 320 克	butter 320 g
水 8000 毫升	water 8000 ml
沙拉油 200 毫升	salad oil 200 ml
白胡椒粉适量	white pepper moderate amount
火腿 600 克	ham 600 g

步骤

1. 将上述所有材料切成丝或片状（0.2厘米厚）。
2. 把沙拉油倒入炒锅加热后，将泡好的干香菇炒香，再放入一半的黄油至溶化后加入洋葱丝炒至透明状。
3. 放入大葱、鲜香菇、口蘑、火腿炒香。
4. 把炒好的汤料倒入水中，加入泡香菇水，大火煮开，转小火，加入鸡粉和白胡椒粉，熬煮20分钟。
5. 关火稍凉，放入果汁机中打碎，即成浓汤。

PROCEDURES

1. Cut all the ingredients into shreds or slices (0.2 cm thick).
2. Put the salad oil into the frying pan, heat the pan, stir-fry the soaked dry mushrooms. Add half of the butter and stir-fry until melted, put in onion shreds and stir-fry until transparent.
3. Add green Chinese onion, fresh mushrooms, apricot mushrooms, ham and stir-fry until fragrant.
4. Pour the cooked broth into water, add the soaked water of dry mushrooms, boil with high flame, then add chicken powder and white pepper powder, simmer for 20 minutes.
5. Turn off the heat, and smash it in a blender after it cools off.

食材	INGREDIENTS
三文鱼 70 克	salmon 70 g
柠檬 1 / 2 个	lemon half
蒜 4 瓣	garlic 4 PCs
黑胡椒 3 克	black pepper 3 g
蘑菇 50 克	mushroom 50 g
百里香 3 克	thyme 3 g
罗勒酱 10 克	basil sauce 10 g
海盐 10 克	sea salt 10 g
白葡萄酒 10 毫升	white wine 10 ml
沙拉油 500 毫升	salad oil 500 ml
橄榄油 10 毫升	olive oil 10 ml
牛奶适量	milk moderate amount

油封三文鱼

Oil-simmered Salmon

步骤

1. 三文鱼用海盐、黑胡椒、柠檬汁、白葡萄酒腌制，腌制后放在50℃的橄榄油中慢慢浸熟，取出放在空盘中待用。
2. 大蒜剁碎，蘑菇切片，将大蒜入油锅爆香，下入蘑菇和百里香，加入牛奶、海盐、黑胡椒调味。
3. 用炒好的蘑菇垫底，三文鱼放在蘑菇上，配以罗勒酱。

PROCEDURES

1. After pickling salmon with sea salt, black pepper, lemon juice and white wine, simmer the salmon in a pan with 50 degree oil slowly, take it out and place it on an empty plate.
2. Mince the garlic, slice the mushroom. Stir-fry the garlic in a pan with oil, put in mushroom, thyme, milk, sea salt and black pepper.
3. Put the cooked salmon on the mushrooms, and match the basil sauce.

小贴士

1. 油封三文鱼的油温不能过高，50℃左右。
2. 三文鱼从油里取出时，小心鱼肉碎烂，一定用吸油纸吸掉多余的油脂。

Tips

1. The temperature of oil to simmer salmon is about 50 degree.
2. When taking the salmon out from the oil, don't make it broken. Make sure to use the oil-absorbing sheet to absorb excess oil.

玉米浓汤

Corn Puree

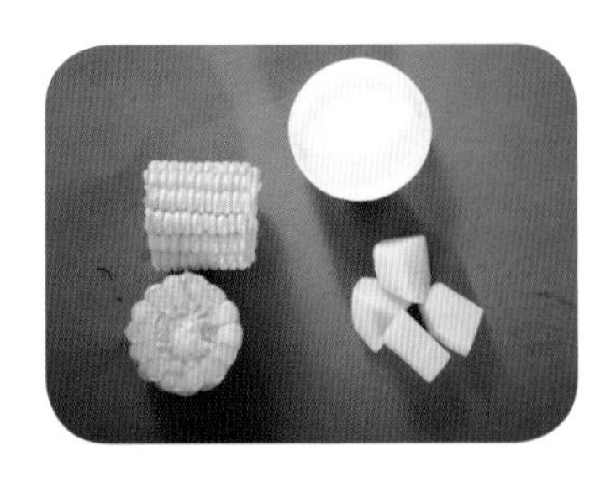

食材	INGREDIENTS
玉米 2 段	corn 2 PCs
土豆 30 克	potato 30 g
淡奶油 30 克	whipping cream 30 g
牛奶 10 毫升	milk 10 ml
海盐 10 克	sea salt 10 g
蔬菜高汤适量	vegetable soup-stock moderate amount

步骤

1. 玉米粒和土豆加蔬菜高汤煮熟。
2. 加牛奶和淡奶油、海盐调味后，放冷却，用搅拌机打成浓汤。

PROCEDURES

1. Cook the corn kernels and potato with vegetable soup-stock.
2. Add milk, whipping cream and sea salt. Smash them in a blender after they cool off.

小贴士

玉米粒打碎时一定要打得很细。

Tips

Corn kernels must be smashed well.

土豆培根浓汤

Potato and Bacon Puree

食材	INGREDIENTS
土豆 2 个	potato 2 PCs
培根 2 片	bacon 2 PCs
淡奶油 30 克	whipping cream 30 g
旱金莲叶 1 片	nasturtium leaf 1 PC
大蒜末适量	garlic pieces moderate amount
海盐 10 克	sea salt 10 g
橄榄油 20 克	olive oil 20 g

步骤

1. 土豆去皮，清洗干净，切小块。
2. 锅内放橄榄油炒香培根，加入土豆、清水熬煮至土豆熟烂。
3. 将淡奶油和海盐调入煮好的土豆汤内，一起用搅拌机打碎成浓汤。
4. 培根切片煎香放在汤上，撒上大蒜末和旱金莲叶子即可食用。

PROCEDURES

1. Peel the potatoes, wash them and cut into small cubes.
2. Pour olive oil into the pan and stir-fry bacons until fragrant, add potatoes, simmer with water until the potatoes are ripe.
3. Put whipping cream and sea salt into the cooked soup, smash them into puree in a blender.
4. Slice the bacon and fry them until fragrant, put them on the soup, sprinkle garlic pieces and nasturtium leaf.

小贴士

选择淀粉含量较高的土豆，会使这道汤更浓稠顺滑。

Tips

Potatoes with high starch will make the puree smooth.

步骤

1. 牛排用海盐、胡椒、红酒、迷迭香、干葱腌制。
2. 热锅冷油煎制牛排，根据自己的口味煎制生熟度，煎好后将一颗大蒜，一枝百里香放入锅中。
3. 加入黄油，离火煎制，煎出大蒜和百里香的香味，煎制时将油不断地淋在牛肉上，煎好后放入盘中1～2分钟后浇上黑松露牛肉酱。
4. 配以手指萝卜、胡萝卜球、芦笋、土豆泥即可食用。

PROCEDURES

1. Marinate the beef with sea salt, pepper, red whine, rosemary and dry green onion.
2. Fry the beef with heated pan cold oil, fry it according to your taste of maturity. Put one garlic and one thyme in the pan after frying.
3. Add butter, fry away from the fire. Fry the garlic and thyme until they are fragrant. Sprinkle the oil on the beef while frying. After frying, put on plate 1 ~ 2 minutes and then pour in black truffle sauce. And sprinkle black truffle beef sauce after it's put on the plate for one or two minutes.
4. Serve with finger radish, carrot ball, asparagus and mashed potatoes.

小贴士

1. 黑松露牛肉酱的制作参考前面部分介绍。
2. 煎牛肉一定要热锅冷油。

Tips

1.Refer to the previous introduction to cook black truffle beef sauce.
2. When the beef is fried, heat the pot first and then put in cold oil.

煎澳洲和牛土豆泥黑松露牛肉酱

Fried Australian Beef Served with Mashed Potato and Black Truffle Beef Sauce

食材

和牛 200 克
胡椒 3 克
海盐 8 克
迷迭香 1 枝
干葱 1 棵
大蒜 1 瓣
百里香 1 枝
黄油 20 克
黑松露牛肉酱适量
红葡萄酒 10 毫升
手指萝卜 1 根
胡萝卜球 2 粒
芦笋尖 1 枝
土豆泥 30 克
沙拉油适量

INGREDIENTS

Australian beef 200 g
pepper 3 g
sea salt 8 g
rosemary 1
dry green onion 1 PC
garlic 1 PC
thyme 1 PC
butter 20 g
black truffle beef sauce moderate amount
red wine 10 ml
finger radish 1 PC
carrot ball 2 PCs
asparagus tip 1 PC
mashed potatoes 30 g
salad oil moderate amount

番茄罗勒冷汤

Tomato and Basil Cold Soup

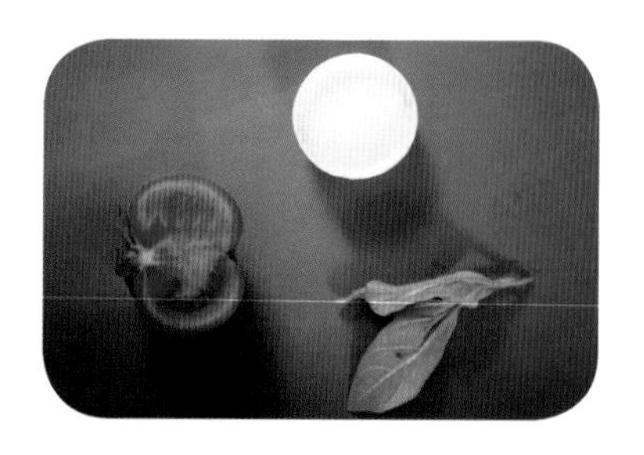

食材	INGREDIENTS
番茄 1 / 2 个	tomato half
新鲜罗勒叶 2 片	fresh basil leaf 2 PCs
橄榄油 10 毫升	olive oil 10 ml
海盐适量	sea salt moderate amount
柠檬汁 10 毫升	lemon juice 10 ml
松子 10 克	pine nut 10 g

步骤

1. 将新鲜番茄清洗干净，去皮，切块。
2. 新鲜罗勒叶清洗干净。
3. 将所有食材混合放入搅拌机内搅打即可。

PROCEDURES

1. Wash and peel the tomato, cut into small cubes.
2. Wash the fresh basil leaves clean.
3. Mix all the ingredients and smash them in a blender.

小贴士

注意汤的浓稠度，调味要适中。

Tips

Pay attention to the thickness of the soup and add less dressing.

橙汁酸菜鸭腿·红酒酱

Duck Leg Cooked with Orange Juice, Pickled Chinese Cabbage and Red Wine Sauce

食材	INGREDIENTS
鸭腿 1 个	duck leg 1 PC
橙子 1 个	orange 1 PC
小洋葱 2 个	small onion 2 PCs
百里香 2 枝	thyme 2 PCs
黑胡椒 5 克	black pepper 5 g
橙子皮 5 克	orange peel 5 g
海盐 5 克	sea salt 5 g
酸菜 20 克	pickled Chinese cabbage 20 g
红酒酱 10 克	red wine sauce 10 g
红葡萄酒 20 毫升	red wine 20 ml

步骤

1. 鸭腿用海盐、黑胡椒、橙子皮、百里香、小洋葱、红葡萄酒腌制好后，放入锅内煎到鸭皮变色，放进烤箱烤熟。
2. 取出烤熟的鸭腿 ，摆上加热的酸菜、橙子，浇上红酒酱。

PROCEDURES

1. Marinate the duck leg with sea salt, black pepper, orange peel, thyme, small onion and red wine. Fry the marinated duck leg until the duck skin turns golden yellow, bake it in an oven.
2. Take duck leg out, match the heated pickled Chinese cabbage, orange and sprinkle the red wine sauce.

小贴士

鸭腿用油煎好后，放烤箱烤熟，如果没有烤箱则小火慢慢煎熟。

Tips

After frying the duck leg, bake it in an oven. If there is no oven, fry it with low flame until it’s ripe.

生蚝土豆慕斯

Oyster and Potato Mousse

食材	INGREDIENTS
生蚝 2 个	oyster 2 PCs
土豆慕斯	potato mousse
青瓜片 2 片	cucumber slices 2 PCs
香叶 1 片	myrcia 1 PC
海盐 3 克	sea salt 3 g
柠檬汁 5 毫升	lemon juice 5 ml
黄油 20 克	butter 20 g
牛奶 20 毫升	milk 20 ml

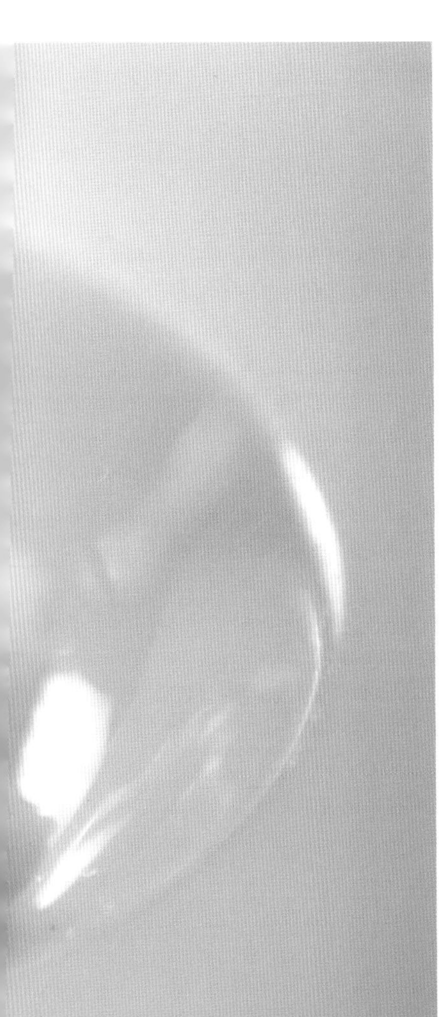

步骤

1. 生蚝撬开取肉，用海盐水浸泡，泡好后放在厨房纸上吸干水分。
2. 土豆用牛奶、黄油、海盐、香叶煮熟。
3. 放入搅拌机打成泥，打好后放入虹吸瓶做成慕斯。
4. 将生蚝、青瓜片放在慕斯上即可。

PROCEDURES

1. Break the oyster for meat, soak it in sea salt water, and blot the extra water with kitchen paper.
2. Boil the potato with milk, butter, sea salt and myrcia.
3. Put the potatoes into the blender for puree, then put the puree into siphon bottle and make mousse.
4. Put the oyster, cucumber slices on the mousse.

小贴士

生蚝一定要加柠檬汁，杀菌。

Tips

Make sure to add lemon juice on the oyster to kill the bacteria.

煎鲈鱼 · 酸黄瓜

Fried Bass and
Sour Cucumber

食材	INGREDIENTS
鲈鱼 80 克	bass 80 g
豌豆 5 根	pea 5 PCs
胡萝卜 1 根	carrot 1
酸黄瓜 1/2 根	pickled cucumber 1/2
西蓝花 1 颗	broccoli 1 PC
海盐 5 克	sea salt 5 g
白葡萄酒 10 毫升	white wine 10 ml
胡椒粉 5 克	pepper powder 5 g
柠檬汁 5 毫升	lemon juice 5 ml
橄榄油 5 毫升	olive oil 5 ml
黄油 10 克	butter 10 g

步骤

1. 鲈鱼用海盐、白葡萄酒、胡椒粉、柠檬汁腌制好放在烧热的锅内，加入橄榄油，煎至鱼皮金黄色，变脆。加入黄油溶化增香。
2. 西蓝花用清水煮熟后切碎。
3. 豌豆、胡萝卜煮好后加白葡萄酒清炒。
4. 将煎好的鲈鱼配以清炒的蔬菜即可。
5. 摆上酸黄瓜。

PROCEDURES

1. Marinate the bass with sea salt, white wine, pepper powder and lemon juice. Fry the marinated bass in a pan until the fish skin turns golden yellow. Add butter to increase the fragrance.
2. Cut the broccoli into pieces after boiling it.
3. Fry the pea and carrot with white wine after boiling them.
4. Put the fried bass and cooked vegetables together.
5. Match the pickled cucumber.

小贴士

鲈鱼煎的时候用厨房纸吸干多余水分，油温要高。（避免鱼肉被煎碎。）

Tips

Blot the extra water on the bass before frying it. The temperature of the oil should be high.

步骤

1. 鹅肝撒少许海盐、胡椒粉、面粉，放在无油的锅中煎制，两面煎至金黄色，喷入苹果醋，可使鹅肝不会过于油腻，待苹果醋挥发取出鹅肝待用。
2. 凤梨切成圆圈，蘸白糖放入平底锅中煎制，煎至白糖变成焦糖色。
3. 喷入白兰地酒，待白兰地酒挥发，取出凤梨，放在盛有鹅肝的盘中。
4. 挤入香橙酱，配以香草苗。

PROCEDURES

1. Sprinkle a pinch of sea salt, pepper powder and flour on the goose liver, fry the liver in a pan without oil until both sides turn golden yellow. Spray some apple cider vinegar which can make goose liver not greasy, take out the liver after the vinegar has volatilized.
2. Cut the pineapple into circles, dip in sugar and fry in a pan until the sugar melts into caramel color.
3. Spray brandy, after it volatilizes, put the pineapple in the plate with goose liver.
4. Add some orange jam, with the vanilla seedling.

小贴士

鹅肝不用解冻直接煎，掌握好火候，不能煎煳，煎熟至两面金黄色。

Tips

Don't defrost the goose live, fry it directly. Take care of the time of frying.

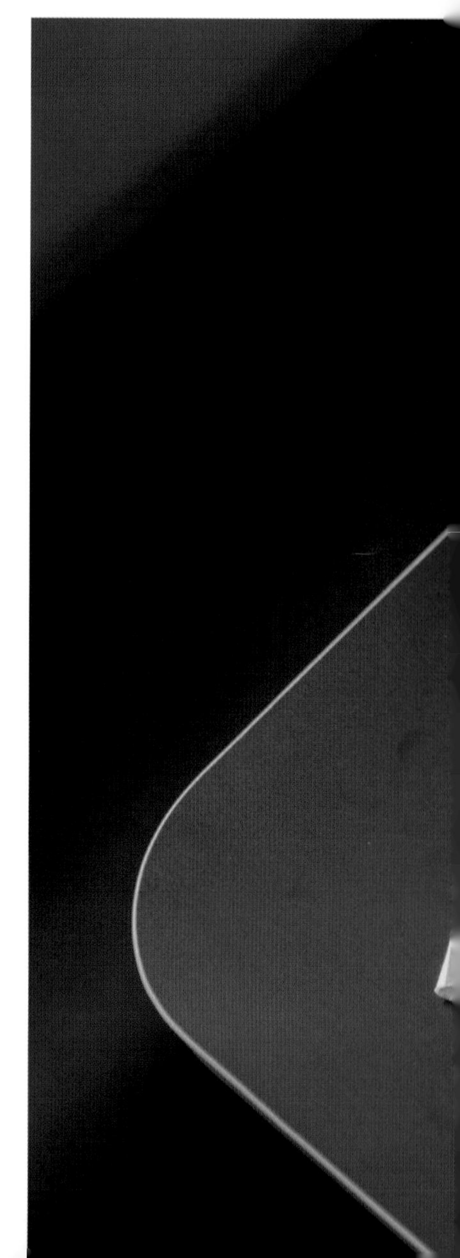

煎鹅肝 · 焦糖凤梨

Fried Goose Liver Served with Caramelized Pineapple

食材

鹅肝 1 片
焦糖凤梨 1 片
香橙酱
海盐 3 克
胡椒粉 2 克
面粉 10 克
苹果醋 10 毫升
白糖 30 克
白兰地酒 5 毫升
香草苗少许

INGREDIENTS

goose liver 1 PC
pineapple 1 slice
orange jam
sea salt 3 g
pepper powder 2 g
flour 10 g
apple cider 10 ml
sugar 30 g
brandy 5 ml
vanilla seedling a few

奶油南瓜浓汤

Cream and Pumpkin Puree

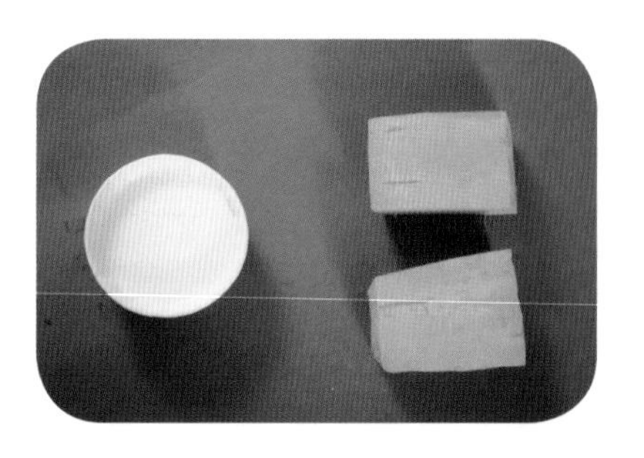

食材	INGREDIENTS
南瓜 200 克	pumpkin 200 g
淡奶油 60 克	whipping cream 60 g
胡萝卜 50 克	carrot 50 g
土豆 50 克	potato 50 g
洋葱 60 克	onion 60 g
水 800 毫升	water 800 ml
沙拉油适量	salad oil moderate amount
黄油 10 克	butter 10 g
海盐少许	sea salt a little

步骤

1. 沙拉油炒香洋葱、土豆、胡萝卜、南瓜，加水熬煮。
2. 加入淡奶油，和黄油、海盐，调味后，放置冷却，用搅拌机搅打成浓汤。

PROCEDURES

1. Stir-fry onion, potatoes, carrots and pumpkins with salad oil until fragrant, add water and boil them for a while.
2. Add whipping cream, butter and sea salt, and cool them for a while.Smash them into puree in a blender.

小贴士

注意汤汁的浓稠度和熬煮时间。

Tips

Pay attention to the thickness of the puree and the time of cooking.

芦笋浓汤

Asparagus Puree

食材	INGREDIENTS
芦笋 500 克	asparagus 500 g
淡奶油 30 克	whipping cream 30 g
海盐 10 克	sea salt 10 g
橄榄油适量	olive oil moderate amount

步骤

1. 芦笋去皮，清洗干净，切小段。
2. 锅内放橄榄油炒香芦笋，加水熬煮至芦笋成熟。
3. 将淡奶油和海盐调入煮好的汤内，一起用搅拌机打碎成浓汤。

PROCEDURES

1. Peel the asparagus, wash it clean and cut it into small segments.
2. Pour olive oil into the pan and stir-fry asparagus until fragrant, add water, simmer until the asparagus is ripe.
3. Put whipping cream and sea salt into the cooked soup, smash them into puree in a blender.

小贴士

芦笋不要煮得过久，否则颜色会变，应保持翠绿的颜色。

Tips

In order to keep the asparagus green, don't cook it too long.

步骤

1. 慕斯层配方中的吉利丁片用冷水浸泡备用。
2. 用模具在海绵蛋糕片上切出蛋糕片，放在模具底部备用。
3. 桃子果茸放入搅拌机打成糊状后隔水加热，将已经泡软的吉利丁片捞出沥干水分后加入果茸糊中搅拌至溶化。
4. 淡奶油加糖打发至六成（浓稠流动状）。
5. 将打好的淡奶油与冷却的果茸糊混合翻拌均匀后即完成慕斯糊。
6. 将慕斯糊装入裱花袋内挤进已经铺好蛋糕底的模具中，不要挤满，留出1厘米左右的高度用来做樱花镜面。
7. 冷藏一晚后，用吹风机在模具周围吹一下，就可以轻松脱模。
8. 将草莓切成两瓣，摆在樱花片周围。

PROCEDURES

1. Soak the gelatine in cold water.
2. Cut the sponge cake into pieces with mould, and put them at the bottom of the mould.
3. Smash the peach fruit puree in a blender and heat it. Take out the soaked gelatine and drain it, put it into fruit paste and stir to melt.
4. Whisk the whipping cream to dense paste.
5. Mix the whisked cream with cooled fruit paste and the mousse is done.
6. Place the mousse paste in the mould bag and squeeze it into a mould that has been covered with cake at the bottom. Don' t fill the mould, leave about 1 cm for the cherry mirror.
7. After a night of refrigeration, blow around the mould with a dryer, and the mould is removed easily.
8. Cut the strawberry into halves, and place them around the cherry blossom.

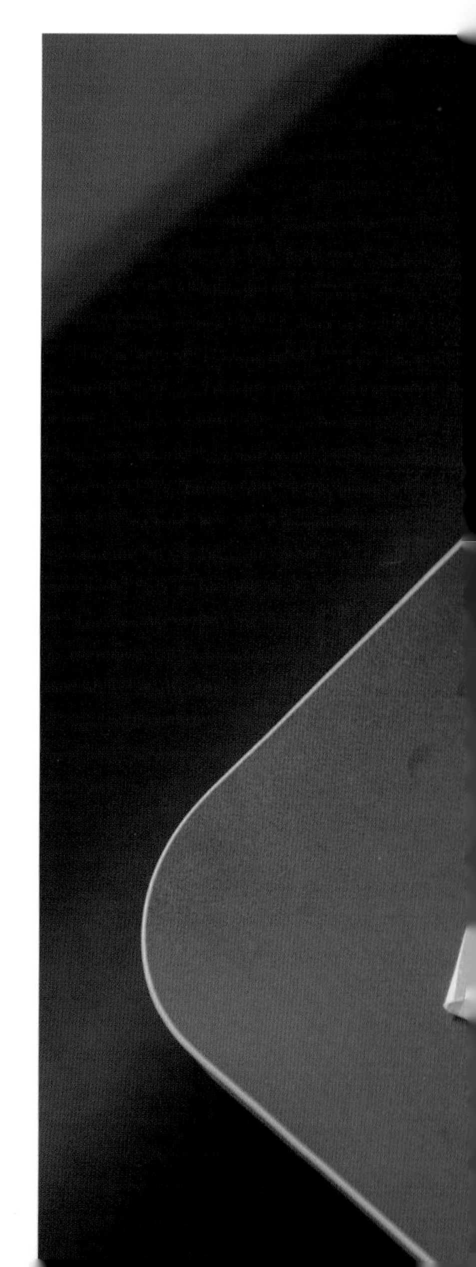

樱花桃子慕斯

Sakura and Peach Mousse

食材

海绵蛋糕底 1 片
慕斯层：
淡奶油 120 克
桃子果茸 120 克
细砂糖 15 克
吉利丁 6 克
草莓 1 个
镜面：
樱花 2 片
水 10 毫升
吉利丁 3 克

INGREDIENTS

sponge cake bottom 1 PC
mousse layers:
whipping cream 120 g
peach fruit puree 120 g
berry sugar 15 g
gelatine 6 g
strawberry 1 PC
cherry mirror:
cherry blossom 2 PCs
water 10 ml
gelatine 3 g

胡萝卜浓汤

Carrot Puree

食材	INGREDIENTS
胡萝卜 2 根	carrot 2 PCs
淡奶油 30 克	whipping cream 30 g
海盐 10 克	sea salt 10 g
橄榄油 10 毫升	olive oil 10 ml

步骤

1. 胡萝卜去皮，清洗干净，切小块。
2. 锅内放橄榄油炒香胡萝卜，加水熬煮至胡萝卜熟烂。
3. 将淡奶油和海盐调入煮好的汤内，一起用搅拌机打碎成浓汤。

PROCEDURES

1. Peel the carrots, wash them clean and cut into small cubes.
2. Pour olive oil into the pan and stir-fry carrots until fragrant, add water, simmer until the carrots are ripe.
3. Put whipping cream and sea salt into the cooked soup, smash them into puree in a blender.

小贴士

注意汤汁的浓稠度。

Tips

Pay attention to the thickness of the puree.

带子薄片
Scallop Slices

食材	INGREDIENTS
带子 2 个	scallop 2 PCs
柠檬 1 / 2 个	lemon half
南瓜子 3 克	pumpkin seeds 3 g
黑胡椒 2 克	black pepper 2 g
可食用的有机花草适量	edible organic plants moderate amount
海盐 5 克	sea salt 5 g
橄榄油 5 毫升	olive oil 5 ml
意大利蜂蜜醋酱	Italian honey vinegar

步骤

1. 带子切成薄片，铺在盘子里。
2. 南瓜子切碎撒在带子上。
3. 将胡椒粉、海盐、橄榄油均匀地撒在上面，配以可食用的有机花草。
4. 滴入少许意大利蜂蜜醋酱，再用柠檬挤出柠檬汁滴在带子上即可。

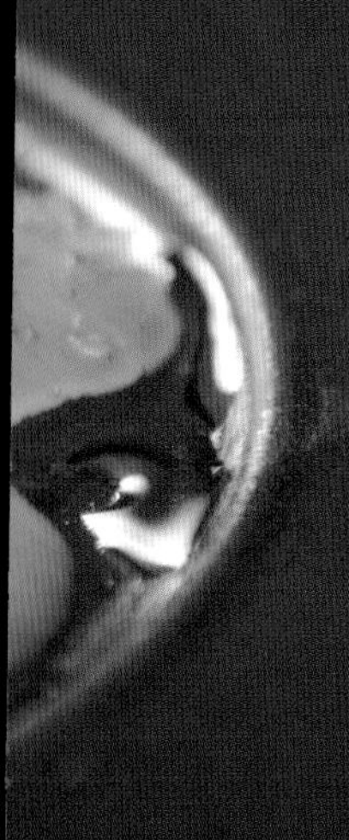

PROCEDURES

1. Slice the scallop and place them on the plate.
2. Chop the pumpkin seeds into pieces and sprinkle them on the scallop
3. Sprinkle black pepper powder, sea salt and olive oil on the scallop, match edible organic plants.
4. Drop a little Italian honey vinegar, squeeze some lemon juice on the scallop.

小贴士

带子要自然解冻，用厨房纸吸干多余的水分，带子生吃，所以一定要选择可以生食的品质。要加柠檬汁，杀菌并增添风味。
注意：生食的食材请用专用的菜板和刀具，避免细菌污染。

Tips

Defrost the scallop naturally. Use kitchen paper to blot the excess water. Choose the scallop that can be eaten raw. Make sure to add lemon juice to kill the bacteria and add flavor.
Notes: In order to avoid bacteria, use special cutting boards and knives to deal with raw ingredients.

蔬菜抄手牛肉清汤

Vegetable Wonton and Beef Soup

食材	INGREDIENTS
西蓝花 10 克	broccoli 10 g
胡萝卜 10 克	carrot 10 g
芹菜适量	celery moderate amount
青豆适量	green bean moderate amount
牛肉清汤	beef soup
海盐 5 克	sea salt 5 g
橄榄油 1 毫升	olive oil 1 ml

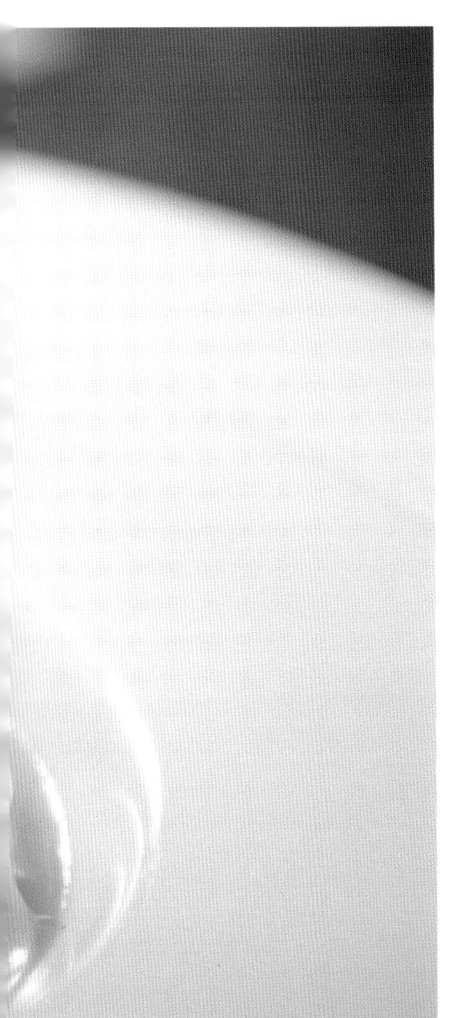

步骤

1. 将胡萝卜、芹菜、青豆分别用水焯一下，捞出，沥水，然后分别剁碎，加海盐、橄榄油，搅拌成馅，用面皮包馅做成蔬菜抄手。
2. 用清水将蔬菜抄手煮熟后捞出。
3. 西蓝花清水煮好。
4. 牛肉清汤煮开后加海盐倒进餐具内，放入蔬菜抄手和西蓝花，滴橄榄油。

PROCEDURES

1. Blanch carrot, celery and green beans, take them out and chop them into pieces. Add sea salt and olive oil, stir them into fillings, and wrap them into vegetable wontons.
2. Boil vegetable wontons in water and then take them out.
3. Boil broccoli in water.
4. Pour the boiled beef soup into a bowl, put vegetable wontons and broccoli, and drop some olive oil.

小贴士

抄手要包好，不要煮破。

Tips

Wrap the vegetable wontons well so that they are not easy to break while cooking.

煎鹅肝·焦糖苹果

Fried Goose Liver Served with Caramelized Apple

食材	INGREDIENTS
鹅肝 1 块	goose liver 1 PC
苹果 1 个	apple 1 PC
香橙酱 5 克	orange jam 5 g
有机新鲜花草	fresh organic plants
白兰地酒少许	brandy a little
海盐 4 克	sea salt 4 g
白糖 20 克	sugar 20 g
胡椒粉 3 克	pepper powder 3 g
面粉 20 克	flour 20 g
苹果醋	apple cider

步骤

1. 鹅肝撒少许海盐、胡椒粉、面粉，放在无油的锅中煎制，两面煎至金黄色，喷入苹果醋，可使鹅肝不会过于油腻，待苹果醋挥发，取出鹅肝待用。
2. 苹果切成圆圈，蘸白糖放入平底锅中煎制，煎至白糖变成焦糖色。
3. 喷入白兰地酒，待白兰地酒挥发，取出苹果，放在盛有鹅肝的盘中。
4. 挤入香橙酱，配以有机新鲜花草。

PROCEDURES

1. Sprinkle a pinch of sea salt, pepper powder and flour on the goose liver, fry the liver in a pan without oil until both sides turn golden yellow. Spray some apple cider vinegar which can make goose liver not greasy, take out the liver after the vinegar has volatilized.
2. Cut the apple into circles, dip in sugar and fry in a pan until the sugar melts into caramel color.
3. Spray brandy, after it volatilizes, put the apple in the plate with goose liver.
4. Add some orange jams, and match fresh organic plants.

煎鳕鱼配奶油烩蘑菇番茄莎莎

Fried Cod Served with Cream-stewed Mushroom and Tomato Salsa

食材	INGREDIENTS
鳕鱼 100 克	cod 100 g
青豆几颗	green bean a few
蘑菇 50 克	mushroom 50 g
百里香 1 枝	thyme 1 PC
迷迭香 1 枝	rosemary 1 PC
番茄 1 / 2 个	tomato half
胡椒粉 4 克	pepper powder 4 g
橄榄油 10 克	olive oil 10 g
淡奶油 10 克	light cream 10 g
海盐 10 克	sea salt 10 g
柠檬汁 4 毫升	lemon juice 4 ml

步骤

1. 青豆用水焯一下，备用。
2. 番茄去皮，切成细丁，用海盐、胡椒粉、橄榄油腌制，制成番茄莎莎。
3. 鳕鱼用海盐、胡椒粉、柠檬汁、橄榄油腌制，腌好后放在锅内煎熟，取出。
4. 用奶油将蘑菇烩熟，铺在盘底，将煎好的鳕鱼放在上面，配以青豆、百里香、迷迭香及做好的番茄莎莎。

PROCEDURES

1. Blanch the green bean for use.
2. Peel the tomato, cut it into small cubes, pickle them with sea salt, pepper powder and olive oil to make tomato salsa.
3. After marinating cod with sea salt, pepper powder, lemon juice and olive oil, fry the cod in a pan until ripe.
4. Stew the mushrooms with cream, and put the cooked mushrooms on the bottom of a plate, put the fried cod on the mushrooms, and serve green bean, thyme, rosemary, tomato salsa.

步骤

1. 豆腐慕斯层配方中的吉利丁片用冷水（配方外）浸泡备用。
2. 内酯豆腐放入搅拌机打成糊状后隔水加热，将已经泡软的吉利丁片捞出沥干水分后加入内酯豆腐糊中搅拌至溶化，淡奶油加糖打发至六成（浓稠流动状），打好的淡奶油与冷却的内酯豆腐糊混合翻拌均匀后即完成慕斯糊。
3. 抹茶慕斯层制作：慕斯层配方中的吉利丁片用冷水浸泡备用。
4. 抹茶粉加牛奶搅拌后隔水加热，将已经泡软的吉利丁片捞出沥干水分后加入搅拌至溶化，淡奶油加糖打发至六成（浓稠流动状），打好的淡奶油与冷却的抹茶糊混合，翻拌均匀后即完成慕斯糊。
   以上两层慕斯依次装进杯子内。

PROCEDURES

1. Soak the gelatine in cold water.
2. Smash the tofu into paste in a blender and heat it. Take out the soaked gelatine and drain it, put it into tofu paste and stir to melt. Whisk the light cream to dense paste, mix the cream with cooled tofu paste.
3. The cooking of matcha mousse: Soak the gelatine in cold water.
4. Stir matcha powder and milk and heat them. Take out the soaked gelatine and drain it, put it into milk and stir to melt. Whisk the light cream and sugar to dense paste, mix the cream with cooled matcha paste. Put the above two layers of mousse into the cup.

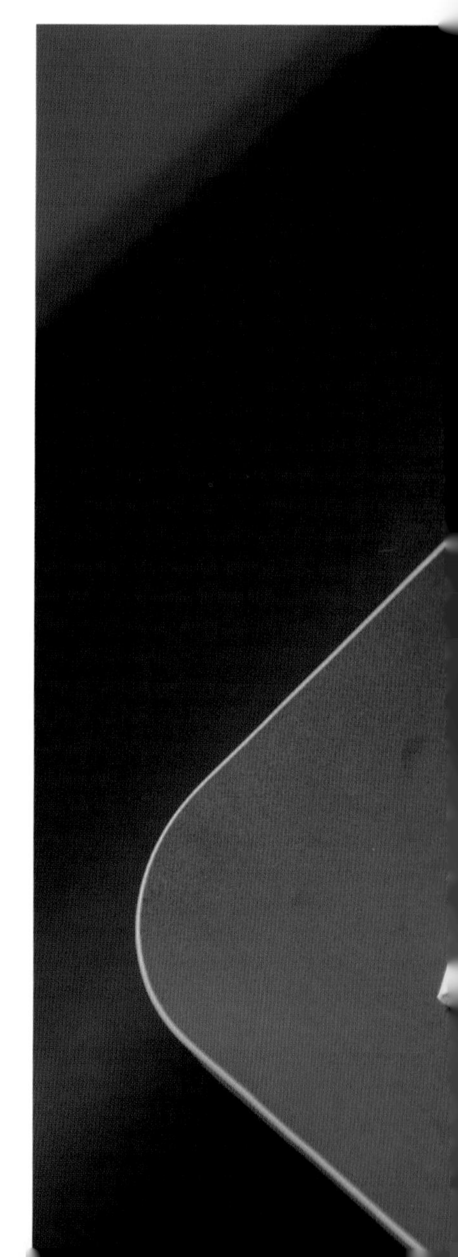

豆腐抹茶慕斯

Tofu and Matcha Mousse

食材

豆腐慕斯层：
淡奶油 120 克
内酯豆腐 120 克
细砂糖 15 克
吉利丁 6 克
抹茶慕斯层：
淡奶油 120 克
牛奶适量
抹茶粉 10 克
细砂糖 15 克
吉利丁 6 克

INGREDIENTS

tofu mousse layer :
light cream 120 g
tofu 120 g
berry sugar 15 g
gelatine 6 g
matcha mousse layer :
light cream 120 g
milk moderate amount
matcha powder 10 g
berry sugar 15 g
gelatine 6 g

培根煎鳕鱼

Bacon and Fried Cod

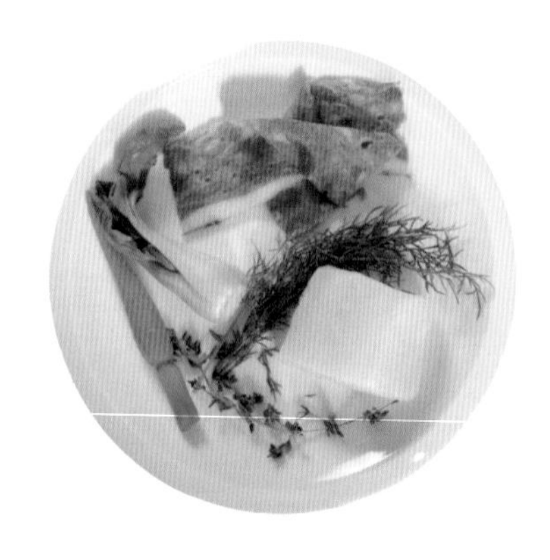

食材	INGREDIENTS
培根 2 片	bacon 2 PC
鳕鱼 80 克	cod 80 g
胡萝卜 1 根	carrot 1 PC
油菜 1 棵	cole 1 PC
百里香 1 枝	thyme 1 PC
迷迭香 1 枝	rosemary 1 PC
花菜泥酱 10 克	cauliflower puree 10 g
海盐 5 克	sea salt 5 g
胡椒粉 3 克	pepper powder 3 g
白葡萄酒 5 毫升	white wine 5 ml
柠檬汁 5 毫升	lemon juice 5 ml
沙拉油适量	salad oil moderate amount

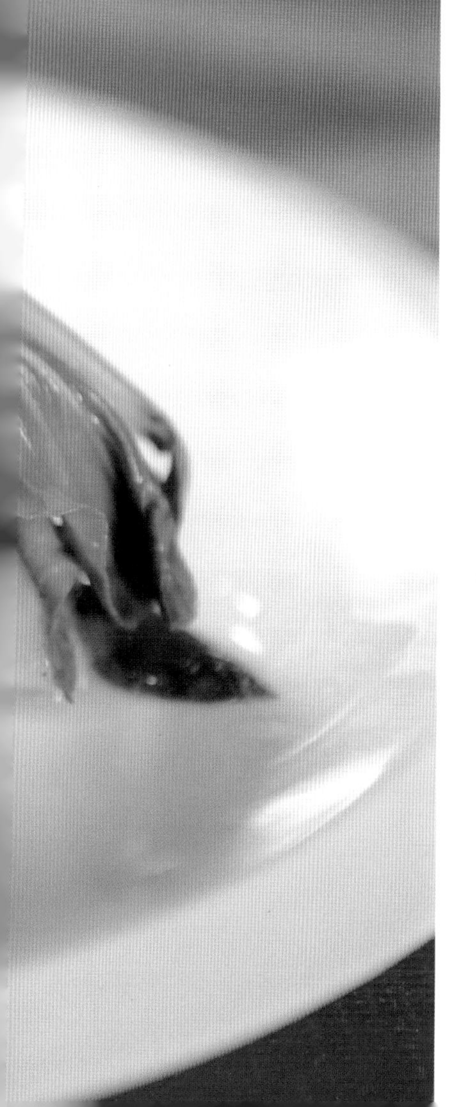

步骤

1. 鳕鱼用海盐、胡椒粉、白葡萄酒、柠檬汁腌好，再用培根包好，放入油锅内用小火慢慢煎制。
2. 将所有的蔬菜煮熟后清炒，炒熟后放入盘中垫底。
3. 将煎好的鳕鱼放在上面，配以做好的花菜泥酱，用百里香、迷迭香摆盘即可。

PROCEDURES

1. Marinate the cod with sea salt, pepper powder, white wine and lemon juice, wrap the marinated cod with bacon, fry the cod with low flame.
2. Boil all the vegetables and stir-fry them, put them on the bottom of a plate.
3. Put the fried cod on the vegetables, add cauliflower puree, and put thyme and rosemary on the plate.

小贴士

培根本身含有海盐分，在腌制鳕鱼时注意海盐的用量，避免过咸。

Tips

Pay attention to the amount of sea salt while the cod is marinated, since bacon is salty.

烩蔬菜

Braised Vegetables

食材	INGREDIENTS
菜心 1 棵	rape center 1 PC
芦笋 2 根	asparagus 2 PCs
胡萝卜 2 根	carrot 2 PCs
茭瓜 1 / 2 个	marrow half
黄油适量	butter moderate amount
淡奶油适量	light cream moderate amount
豆角 50 克	long bean 50 g
海盐 5 克	sea salt 5 g
橄榄油 10 毫升	olive oil 10 ml
白葡萄酒 10 毫升	white wine 10 ml

步骤

1. 蔬菜清洗干净，可以按照自己喜欢的形状进行处理。
2. 锅内烧热水，下入清洗过的蔬菜煮熟。
3. 锅内倒入橄榄油，溶化的黄油，放入蔬菜，再加入海盐、白葡萄酒、淡奶油调味，煮到白葡萄酒挥发出香味即可。

PROCEDURES

1. Wash the vegetables, and shape them according to one’s favor.
2. Heat water in the pot and boil the ingredients in the pot.
3. Pour olive oil and melted butter into the pot, put in vegetables , sea salt, white wine and light cream, cook until you can smell the flavor of the white wine.

小贴士

各类蔬菜按照质地进行先后顺序煮，避免有的已经熟透，有的半熟。

Tips

Cook the vegetables according to their quality of becoming ripe.

步骤

1. 大麦仁煮熟。
2. 胡萝卜、荷兰豆清洗干净。
3. 锅内放油，再放入大蒜、洋葱炒香，把煮好的大麦仁放入锅中，再放入胡萝卜、荷兰豆、海鲜、白葡萄酒、适量的水，放奶油，慢慢地焖煮，煮到汤汁浓稠时放入海盐调味，起锅即可。

PROCEDURES

1. Boil the barley kernel.
2. Wash the carrot and Holland beans.
3. Pour oil into the pan, put in garlic and onion, stir-fry until fragrant. Add boiled barley kernel, carrot, Holland beans, seafood, white wine, some water and cream, stew slowly, and add some sea salt when the soup is thick, finally place them on the plate.

小贴士

大麦仁提前煮好，海鲜食材提前清洗干净，蛤蜊比较多沙。

Tips

Boil the barley kernel in advance. Wash the seafood in advance. Pay attention to the sands in clams.

海鲜烩饭

Seafood-braised Rice

食材

梭子蟹 1 只
基围虾 2 只
蛤蜊 5 个
胡萝卜 1 根
荷兰豆若干
洋葱 20 克
大麦仁 100 克
白葡萄酒 10 毫升
海盐 8 克
奶油 10 克
蒜 1 瓣
沙拉油适量

INGREDIENTS

swimming crab 1 PC
small-sized shrimp 2 PCs
clam 5 PCs
carrot 1 PC
A number of Holland bean
onion 20 g
barley kernel 100 g
white wine 10 ml
sea salt 8 g
cream 10 g
garlic 1 clove
salad oil moderate amount

茄汁青口贝烩意大利面

Spaghetti Braised with Eggplant Juice and Mussel

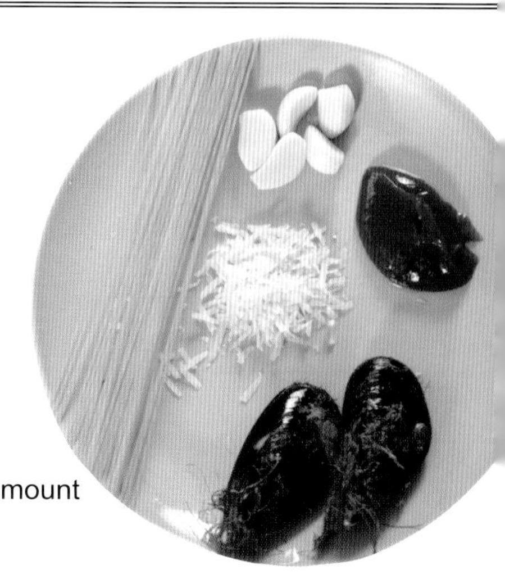

食材	INGREDIENTS
意大利面 100 克	Spaghetti 100 g
大蒜 5 瓣	garlic 5 g
芝士 20 克	cheese 20 g
番茄酱 50 克	tomato sauce 50 g
青口贝 50 克	mussel 50 g
橄榄油 10 毫升	olive oil 10 ml
海盐 5 克	sea salt 5 g
洋葱适量	onion moderate amount
茄汁适量	eggplant sauce moderate amount
沙拉油适量	salad oil moderate amount
白葡萄酒少许	white wine a little

步骤

1. 青口贝用沸水煮开，去掉沙石杂质。
2. 意大利面煮熟。
3. 锅内放沙拉油，放入大蒜、洋葱炒香，加入茄汁、煮好的青口贝、意大利面，加少量的水、白葡萄酒、海盐调味，炒好放入盘中即可。

PROCEDURES

1. Boil the mussels in hot water, remove the impurities.
2. Boil the Spaghetti.
3. Pour salad oil into the pan, put in garlic and onion, stir-fry until fragrant. Add eggplant juice, cooked mussels, Spaghetti, some water, white wine and sea salt, stir-fry and place them on the plate.

小贴士

出锅时撒一些法香碎。

Tips

Sprinkle some thyme when the dish is done.

奶油三文鱼意面

Creamed Salmon Spaghetti

食材	INGREDIENTS
意大利面 100 克	Spaghetti 100 g
三文鱼 50 克	salmon 50 g
西蓝花 1 颗	broccoli 1 PC
洋葱 20 克	onion 20 g
大蒜 3 瓣	garlic 3 PCs
奶油 40 克	cream 40 g
大葱适量	green Chinese onion moderate amount
海盐 8 克	sea salt 8 g
橄榄油 10 毫升	olive oil 10 ml
黑胡椒适量	black pepper moderate amount
柠檬汁适量	lemon juice moderate amount
白葡萄酒少许	white wine a little

步骤

1. 三文鱼用黑胡椒、柠檬汁、海盐、白葡萄酒腌制，腌好后将三文鱼煎熟。
2. 西兰花用热水焯熟。
3. 锅内放入洋葱、大葱段、大蒜炒香，加意大利面，炒熟后再加入少许奶油，收干汤汁，焯熟的西蓝花摆盘。
4. 将煎好的三文鱼放在炒好的意大利面上即可。

PROCEDURES

1. Marinate the salmon with black pepper, lemon juice, sea salt and white wine, then fry the salmon.
2. Blanch the broccoli.
3. Put onion , green Chinese onion segments and garlic into the pan, stir-fry until fragrant. Put the Spaghetti into the pan and stir-fry, add some cream and stew for minutes and place the dish on the plate with broccoli.
4. Place the fried salmon on the cooked Spaghetti.

小贴士

意面提前煮好，煮时加入三文鱼，加些白酒增加味道。

Tips

Boil the Spaghetti with salmon in advance. Add some white wine to add flavor during the boiling.

番茄蛤蜊意面

Tomato and Clam Spaghetti

食材	INGREDIENTS
意大利面 100 克	Spaghetti 100 g
蛤蜊 100 克	clam 100 g
番茄 50 克	tomato 50 g
黄油 20 克	butter 20 g
大蒜若干	garlic a little
芝士 20 克	cheese 20 g
法香碎 5 克	thyme 5 g
洋葱适量	onion moderate amount
海盐 8 克	sea salt 8 g
橄榄油 10 毫升	olive oil 10 ml
白葡萄酒少许	white wine a little

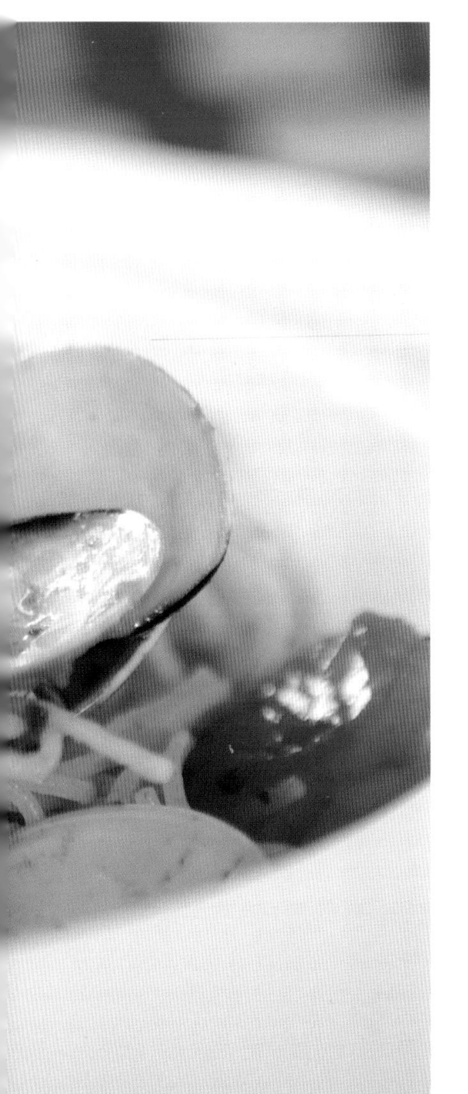

步骤

1. 番茄切小块，大蒜切碎，洋葱切碎，法香切碎。
2. 意大利面煮15分钟捞出，蛤蜊用水煮开。
3. 锅内放橄榄油和黄油，加入洋葱碎、大蒜碎炒香，放入蛤蜊、番茄炒香后加入意大利面，再加入海盐、芝士碎、白葡萄酒调味后起锅即可。

PROCEDURES

1. Cut the tomato into small cubes, chop the garlic, onion and thyme into pieces.
2. Boil the Spaghetti for 15 minutes and take it out, boil the clam.
3. Put olive oil, butter, onion pieces and garlic pieces into the pot, stir-fry until fragrant. Put in clam and tomato, and stir-fry, add Spaghetti, sea salt, cheese pieces and white wine.

小贴士

意面不要煮得过软。

Tips

Don’t boil the Spaghetti too soft.

蔬菜烩饭

Vegetable-braised Rice

食材	INGREDIENTS
大麦仁 50 克	barley kernel 50 g
洋葱 10 克	onion 10 g
西蓝花 10 克	broccoli 10 g
番茄 10 克	tomato 10 g
花菜 10 克	cauliflower 10 g
黄油 10 克	butter 10 g
奶油 10 克	cream 10 g
青豆 10 克	green bean 10 g
海盐适量	sea salt moderate amount
白葡萄酒少许	white wine a little

步骤

1. 大麦仁煮熟，取出。
2. 锅内加入黄油，放入洋葱及蔬菜炒香。
3. 把煮好的大麦仁放入锅中，再放入奶油、海盐、白葡萄酒，烧到汁干，出锅即可。

PROCEDURES

1. Boil the barley kernel and take them out.
2. Put butter, onion pieces and vegetables into the pan and stir-fry until fragrant.
3. Put the boiled barley kernel into the pan, put in cream, sea salt and white wine, stew until the juice becomes dry.

小贴士

大麦仁一定要煮到全部裂开，冲冷水后再用。

Tips

Boil the barley kernel until it is cracked, and spray it with cold water.

果木煎金枪鱼

Tuna Fried With Fruitwood

食材	INGREDIENTS
金枪鱼 100 克	tuna 100 g
西洋菜 1 片	lettuce leaf 1 PC
茭瓜 50 克	marrow 50 g
黑胡椒粒 20 克	black pepper 20 g
柠檬 1 / 2 个	lemon half
海盐 10 克	sea salt 10 g
胡椒粉 2 克	pepper powder 2 g
白葡萄酒 5 毫升	white wine 5 ml
沙拉油适量	salad oil moderate amount

步骤

1. 金枪鱼用胡椒粉、柠檬汁、海盐、白葡萄酒腌制。
2. 腌制过后，加粗黑胡椒粒，热锅冷油煎制，煎至四周变金黄色取出。
3. 煎好的金枪鱼切片、摆盘，配以茭瓜、西洋菜。

PROCEDURES

1. Marinate the tuna with black pepper, lemon juice, sea salt and white wine.
2. Add black pepper grains, fry the tuna with heat pan and cold oil until golden yellow, and then take the tuna out.
3. Slice the tuna, place it on the plate and match marrow and lettuce leaf.

金枪鱼煎制时间不能太久。

Don't fry tuna too long.

白葡萄酒炖牛肉

Beef Stewed with White Wine

食材	INGREDIENTS
牛肉 100 克	beef 100 g
蘑菇 30 克	mushroom 30 g
胡萝卜 20 克	carrot 20 g
土豆 20 克	potato 20 g
洋葱 20 克	onion 20 g
迷迭香 1 枝	rosemary 1 PC
胡椒适量	pepper moderate amount
白葡萄酒 100 毫升	white wine 100 ml
海盐 20 克	sea salt 20 g
大蒜适量	garlic moderate amount
沙拉油适量	salad oil moderate amount

步骤

1. 将牛肉切块，所有蔬菜切块。
2. 锅内放油，将洋葱、大蒜、迷迭香、胡椒放入锅内炒香，放入牛肉，倒入白葡萄酒，加入蔬菜块，再加水煮熟，最后加入海盐调味即可。

PROCEDURES

1. Dice the beef and vegetables.
2. Put oil into the pan, add onion, garlic, rosemary and pepper, stir-fry until fragrant. Put in beef, white wine, vegetables and water, stew them until ripe, finally add some sea salt.

小贴士

牛肉一定要炖软烂。

Tips

The beef must be cooked soft and tender.

鸡肉卷 · 烤番茄

Chicken Roll Served with Baked Tomato

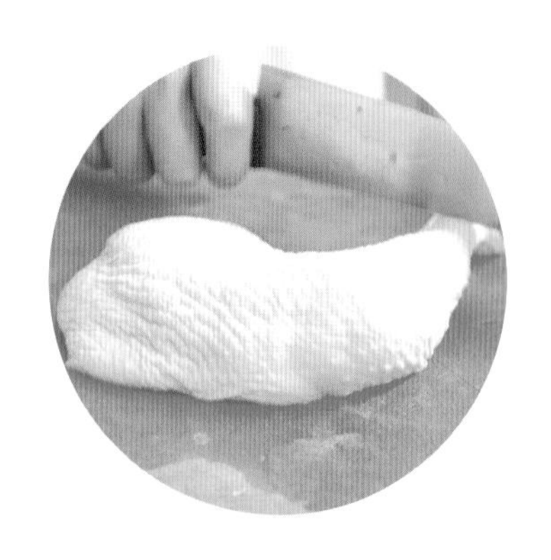

食材 INGREDIENTS

食材	INGREDIENTS
鸡腿 1 个	chicken leg 1 PC
番茄 30 克	tomato 30 g
迷迭香 1 枝	rosemary 1 PC
胡椒 10 克	pepper powder 10 g
海盐 10 克	sea salt 10 g
红酒 10 克	red wine 10 g
橄榄油 20 毫升	olive oil 20 ml
南洋辣酱 20 克	Nanyang chilli sauce 20 g
豆角 20 克	long bean 20 g

步骤

1. 鸡腿去骨，将鸡腿肉切成薄厚均匀的片，用胡椒、海盐、迷迭香、红酒腌制。
2. 腌好后将鸡肉做成卷状，用保鲜膜定型，放在水中煮熟。
3. 煮熟后去掉保鲜膜放在油锅中，煎至金黄色，取出切断摆盘，配以烤番茄和豆角，浇南洋辣酱即可。

PROCEDURES

1. Remove the bones of chicken leg, slice the chicken and marinate with pepper, sea salt, rosemary and red wine.
2. Roll the chicken and mould with plastic wrap, boil it in the water.
3. Remove the plastic wrap and fry in the pan with oil until the chicken turns golden yellow. Cut the chicken into segments and place them on the plate, add baked tomato, long bean and Nanyang chilli sauce.

小贴士

鸡腿去骨后，用海盐、胡椒、红酒、迷迭香腌制，再做成肉卷，用保鲜膜包裹一定要注意成型。
煮熟后取出鸡肉卷再下锅煎，煎到表皮脆，呈金黄色。

Tips

After removing the bones of the chicken leg, marinate with sea salt, pepper, red wine and rosemary, and roll the chicken. Make sure to shape with plastic wrap.
Take out the chicken rolls after they are ripe. Fry them until the skin is golden yellow.

迷迭香土豆煎鸡腿

Fried Chicken Leg Cooked with Rosemary and Potato

食材	INGREDIENTS
土豆 20 克	potato 20 g
鸡腿 1 个	chicken leg 1 PC
迷迭香 1 枝	rosemary 1 PC
海盐 5 克	sea salt 5 g
胡椒粉 10 克	pepper powder 10 g
松露牛肉汁 20 毫升	truffle beef sauce 20 ml

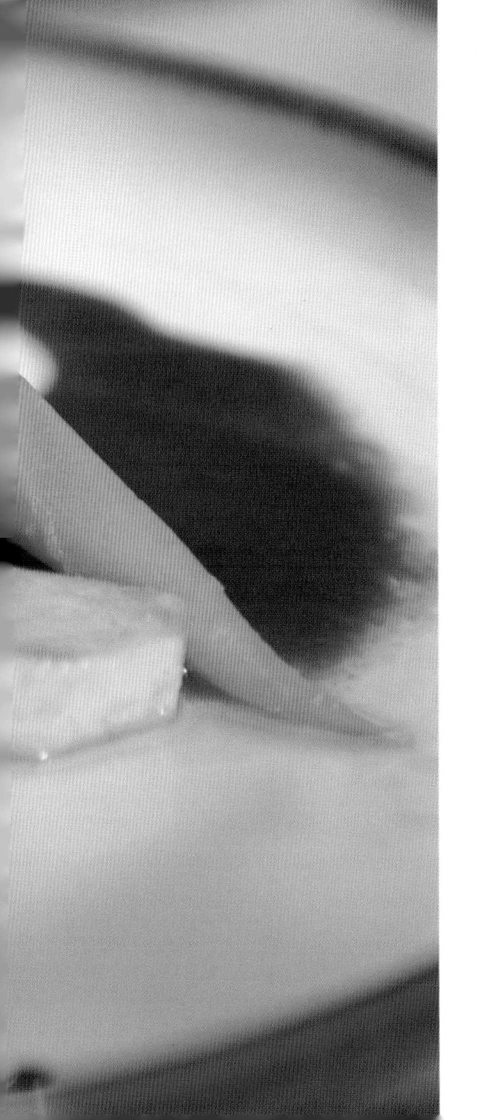

步骤

1. 鸡腿用海盐、胡椒粉、迷迭香腌制好，直接入烤箱180℃烤到表面金黄色。
2. 土豆清洗干净，去皮切成块，撒上海盐、胡椒粉入烤箱烤熟。
3. 土豆和鸡腿都烤好后，依次摆进盘内，浇上松露牛肉汁。

PROCEDURES

1. After marinating chicken leg with sea salt, pepper powder and rosemary, put it into a 180 degree oven and bake until the skin turns golden yellow.
2. Peel the potato and cut it into cubes, sprinkle sea salt and pepper, bake in an oven.
3. Place the baked potato and chicken leg on the plate, and sprinkle truffle beef sauce.

小贴士

鸡腿烤到鸡皮呈金黄色即可，这时的口感是最香脆的。

Tips

Bake the chicken leg until the skin turns golden yellow.

辣椒粉迷迭香羊排

Lamb Chop Cooked with Chilli Powder and Rosemary

食材 INGREDIENTS

食材	INGREDIENTS
新西兰羊排 1 条	New Zealand lamb chop 1 PC
红椒粉 20 克	red pepper powder 20 g
胡椒粉 10 克	pepper powder 10 g
海盐 5 克	sea salt 5 g
烤小蘑菇 10 克	baked small mushroom 10 g
迷迭香 1 枝	rosemary 1 PC
烤红菜头 3 块	beet 3 PCs
茭瓜 50 克	marrow 50 g
红酒 10 毫升	red wine 10 ml
土豆泥 20 克	mashed potato 20 g

步骤

1. 羊排取里脊肉，放入海盐、胡椒粉、红酒腌制。
2. 腌好后在羊排上均匀地抹上红椒粉，用保鲜膜包好，放在水中煮熟。
3. 煮好后去掉保鲜膜切成块，摆盘，浇土豆泥，配迷迭香、小蘑菇、红菜头、茭瓜即可。

PROCEDURES

1. Marinate tenderloin of lamb chop with sea salt, pepper powder and red wine.
2. Cover the marinated lamb chop with red pepper powder, pack it with plastic wrap, and boil it in water.
3. Remove the plastic wrap after boiling, cut the tenderloin into cubes, place them on the plate, and sprinkle mashed potato, match rosemary, small mushroom, beet and marrow.

小贴士

羊排煮的时间为 5 分钟，这样肉质更棒。

Tips

Boil the lamb chop for 5 minutes to get better mutton.

步骤

1. 锅内放油，将大蒜、洋葱切碎，放入锅中，加豆瓣酱炒香，加入蔬菜高汤，煮沸后放入大虾煮熟。
2. 将罗勒叶切丝备用。
3. 煮熟后放入罗勒叶丝，加入海盐、白葡萄酒、柠檬汁调味，放入老抽调色，汤汁快要收干时用淀粉勾芡。倒入锅中，起锅摆盘即可。

PROCEDURES

1. Pour oil into the pan, put in garlic pieces, onion pieces and broad-bean sauce, stir-fry until fragrant. Add vegetable soup-stock, cook the prawn after the soup-stock boils.
2. Chop the basil leaves into shreds.
3. Put in basil shreds, sea salt, white wine, lemon juice and dark soy sauce, add starch to make the soup dense.Place the dish on the plate after cooking.

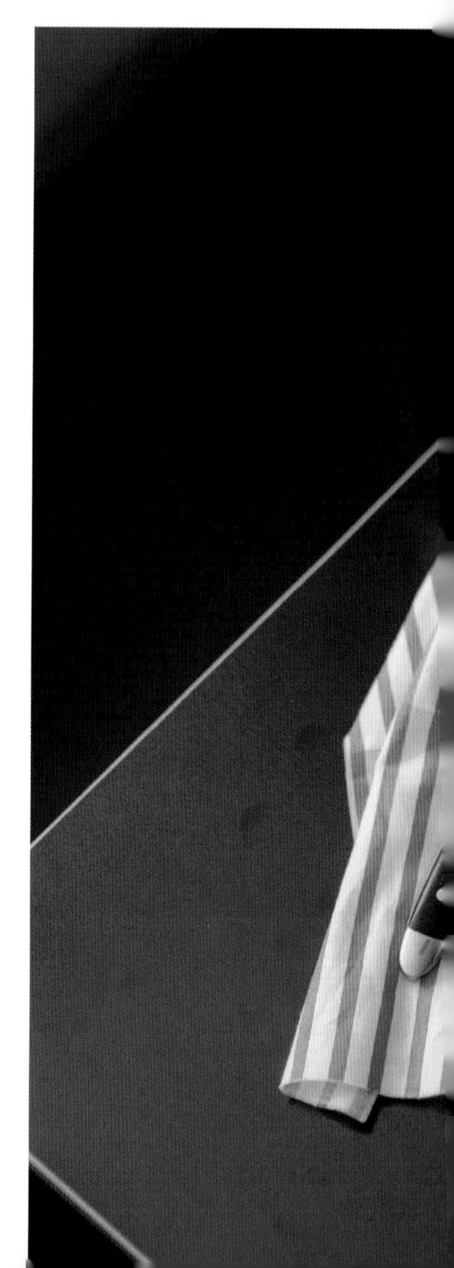

小贴士

蒜蓉豆瓣酱比较咸，可以放一些白糖调和。

Tips

The garlic broad-bean sauce is a little salty, some sugar can be added to mix the salty sauce.

豆瓣酱罗勒烧大虾

Prawn Braised with Broad-bean Sauce and Basil

食材

蒜蓉豆瓣酱 20 克
新鲜罗勒叶 5 克
大虾 300 克
大蒜 5 克
洋葱 10 克
橄榄油 10 毫升
白葡萄酒 10 毫升
柠檬汁 5 毫升
淀粉 5 克
海盐少许
老抽少许
蔬菜高汤适量

INGREDIENTS

garlic broad-bean sauce 20 g
fresh basil leaves 5 g
prawn 300 g
garlic 5 g
onion 10 g
olive oil 10 ml
white wine 10 ml
lemon juice 5 ml
cornstarch 5 g
sea salt a little
dark soy sauce a little
vegetable soup-stock moderate amount

粗盐烤红菜头

Beet Baked with Coarse Salt

食材 INGREDIENTS

食材	INGREDIENTS
红菜头 1 个	beet 1 PC
粗盐 1 袋	coarse salt 1 bag
海盐适量	sea salt moderate amount
胡椒粉少许	pepper powder a little
橄榄油适量	olive oil moderate amount

步骤

1. 红菜头洗干净，放在锡纸内。
2. 烤盘中铺满粗盐，将红菜头放在烤盘内，入烤箱180℃烤1小时左右。
3. 取出红菜头，去锡纸，将红菜头去皮、切块，放少量的海盐、胡椒粉、橄榄油，搅拌均匀即可。

PROCEDURES

1. Wash the beet and put it in tin foil.
2. Cover the bakeware with coarse salt, and place the beet, bake in a 180 degree oven for an hour.
3. Take out the beet and remove the tin foil, peel the beet and cut it into cubes. Add a little sea salt, pepper powder and olive oil, mix well.

小贴士

红菜头清洗干净，用锡纸包裹好，粗盐覆盖入烤箱烤。

Tips

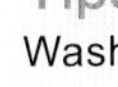

Wash the beet clean, wrap it with tin foil, cover it with coarse salt and bake it in an oven.

白酒烩青口贝

Liquor-braised Mussel

食材	INGREDIENTS
青口贝 100 克	mussel 100 g
胡萝卜 1 根	carrot 1 PC
洋葱 20 克	onion 20 g
芹菜 1 根	celery 1 PC
大蒜 2 瓣	garlic 2 PCs
小洋葱 1 个	small onion 1 PC
大葱 1 根	green Chinese onion 1 PC
橄榄油 10 毫升	olive oil 10 ml
胡椒粉 5 克	pepper powder 5 g
海盐 10 克	sea salt 10 g
柠檬汁 5 毫升	lemon juice 5 ml
白葡萄酒 50 毫升	white wine 50 ml

步骤

1. 青口贝用沸水煮开，去掉沙石杂质。
2. 汤锅中放油、胡椒粉，再放入芹菜、大蒜、洋葱、小洋葱、大葱、胡萝卜炒香。
3. 放入煮好的青口贝，再调入白葡萄酒、柠檬汁。
4. 加清水、海盐熬煮收浓汤汁即可。

PROCEDURES

1. Boil the mussels with hot water, remove the impurity.
2. Put oil, pepper power, celery, garlic, onion, small onion, green Chinese onion and carrot, stir-fry until fragrant.
3. Put in boiled mussels, whine wine and lemon juice.
4. Add water, sea salt, and boil for a while.

小贴士

青口贝应小火慢慢烩煮，让汁液浸进青口贝内，味道更鲜美。

Tips

Cook the mussels with low flame to let the soup immerse into the mussels. It will taste better.

煎牛柳

Fried Beef Fillet

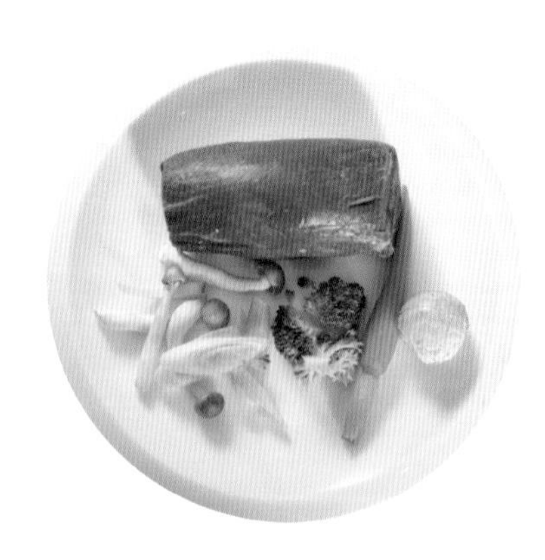

食材 INGREDIENTS

菲力牛排 100 克	filet steak 100 g
小蘑菇 20 克	small mushroom 20 g
胡萝卜 1 根	carrot 1 PC
西蓝花 2 棵	broccoli 2 PCs
孢子甘蓝 1 个	spore olive 1 PC
大葱少许	green Chinese onion a little
茴香头 20 克	fennel bulb 20 g
牛肉汁	beef broth
海盐少许	sea salt a little
胡椒粉少许	pepper powder a little
迷迭香 1 枝	rosemary 1 PC
红酒少许	red wine a little

步骤

1. 牛排用海盐、胡椒粉、迷迭香、红酒腌制。胡萝卜、小蘑菇、西蓝花、孢子甘蓝、茴香头煮熟。
2. 大葱煎至金黄色。
3. 牛排腌好后放在锅中煎到自己喜爱的成熟度，搭配以上煮好的蔬菜、大葱，配上牛肉汁即可。

PROCEDURES

1. Marinate the beef with sea salt, pepper power, rosemary and red wine. Boil carrot, small mushroom, broccoli, spore olive and fennel bulb to ripe.
2. Fry the green Chinese onion into golden yellow.
3. Fry the marinated beef according to personal taste, serve with the boiled vegetables, green Chinese onion and beef broth.

小贴士

牛排煎的时间根据自己喜爱掌握成熟度。

Tips

The time of frying the beef depends on one's taste.

炒鸡蛋

Scrambled Eggs

食材	INGREDIENTS
鸡蛋 4 个	egg 4 PCs
淡奶油 40 克	light cream 40 g
海盐 5 克	sea salt 5 g
培根 2 片	bacon 2 PCs
小蘑菇 20 克	small mushroom 20 g
黄油 10 克	butter 10 g

步骤

1. 平底锅烧开水，将鸡蛋打入盆中，再打入等量的淡奶油。
2. 撒入海盐、黄油，顺时针搅拌，放在热水上，隔水加热。
3. 炒到鸡蛋凝固取出，放入盘中，撒上培根碎和烤好的蘑菇。

PROCEDURES

1. Boil water in a pan, break the eggs into a basin, put in the same amount of light cream into the basin.
2. Put in sea salt and butter, stir clockwise, put the basin on hot water and heat the basin.
3. Stir-fry the eggs, put them into a plate, sprinkle bacon pieces and baked mushrooms.

小贴士

鸡蛋隔水炒到凝固，这样的鸡蛋更滑嫩。

Tips

Stir-fry the eggs to solidification, so that the eggs are tender.

烤澳洲和牛

Baked Australian Beef

食材	INGREDIENTS
和牛 300 克	beef 300 g
茭瓜 100 克	marrow 100 g
迷迭香 1 枝	rosemary 1 PC
海盐 5 克	sea salt 5 g
胡椒碎 3 克	pepper piece 3 g
胡萝卜 1 根	carrot 1 PC
孢子甘蓝 1 个	spore olive 1 PC
橄榄油 10 毫升	olive oil 10 ml
红酒 5 毫升	red wine 5 ml
黄油 20 克	butter 20 g

步骤

1. 澳洲和牛用粗黑胡椒碎、海盐、红酒腌制。
2. 腌好后放入橄榄油煎至两面金黄，放上迷迭香、黄油，进烤箱180℃烤3分钟。胡萝卜和孢子甘蓝用水煮熟。
3. 茭瓜切片，放在锅内两面煎熟取出，摆盘，配上海盐和橄榄油即可。

PROCEDURES

1. Marinate Australian beef with coarse black pepper piece, sea salt and red wine.
2. Fry the beef in a pan with oil until both sides turn golden yellow. Put in rosemary and butter, then bake beef in a 180 degree oven for three minutes. Boil carrot and spore olive to ripe.
3. Slice the marrow, take the slices out after frying them in a pan. Put them on a plate, add sea salt and olive oil.

小贴士

和牛烤的时间不宜过长，五成熟口感更好。

Tips

Don't bake beef too long. Medium-rare tastes better.

大麦腊肉烩饭

Rice Braised with Barley and Preserved Meat

食材	INGREDIENTS
大麦仁 100 克	barley kernel 100 g
腊肉 20 克	preserved meat 20 g
蔬菜丁 50 克	vegetable cubes 50 g
淡奶油 20 克	light cream 20 g
海盐 5 克	sea salt 5 g
黄油 10 克	butter 10 g

步骤

1. 腊肉切成小丁，用清水煮一下，去除肉的咸味。
2. 大麦仁用清水煮至发软，取出待用。
3. 锅内下黄油、蔬菜丁、煮过的腊肉丁，炒香，放入大麦仁，加入适量的水、淡奶油小火慢煮，至液体呈浓稠状，再加入淡奶油，搅拌后即可。

PROCEDURES

1. Cut the preserved meat into small cubes, boil the cubes with clear water to remove the saltiness of the meat.
2. Boil the barley kernel in clear water until it is soft, take the barley kernel out for use.
3. Put butter into the pan, put in vegetable cubes and preserved meat, stir-fry until fragrant. Put in barley kernel, water and light cream, simmer until the soup becomes dense. Add light cream and stir.

小贴士

大麦仁可以提前煮软。

Tips

Boil the barley kernel soft in advance.

竹笋培根烧海参

Sea Cucumber Braised with Bamboo Shoot and Bacon

食材	INGREDIENTS
海参 1 条	sea cucumber 1 PC
培根 30 克	bacon 30 g
竹笋 80 克	bamboo shoot 80 g
油 20 毫升	oil 20 ml
姜 5 克	ginger 5 g
葱 5 克	green onion 5 g
蒜 5 克	garlic 5 g
老抽 3 毫升	dark soy sauce 3 ml
花椒油 1 克	pepper oil 1 g
辣椒油 3 克	chilli oil 3 g
淀粉 5 克	cornstarch 5 g
海盐少许	sea salt a little
香油少许	sesame oil a little

步骤

1. 海参用水发好，新鲜竹笋切成细丁，用水煮过；培根切成细丁。
2. 锅内入油，下入培根、葱、姜、蒜末，炒香；倒入竹笋、清水烧制，要没过全部食材，待食材烧熟，加入老抽、花椒油、辣椒油、海盐调味，然后加入海参一起烧制。
3. 出锅前勾少许芡，加入香油，出锅摆盘即可。

PROCEDURES

1. Soak the sea cucumber in water.Cut the fresh bamboo shoots into small cubes and boil them in water.Cut bacon into small cubes.
2. Put oil into the pan, put in bacon, green onion pieces, ginger pieces and garlic pieces, stir-fry until fragrant. Add bamboo shoots and enough water to cook. After they are ripe, add dark soy sauce, pepper oil, chilli oil and sea salt, finally put in sea cucumber to stew.
3. Use cornstarch to make the soup dense, add some sesame oil and place the dish on a plate.

小贴士

海参需要提前三天用水发好。配以新鲜花草摆盘。

Tips

Soak the sea cucumber in water 3 days in advance. Place fresh flowers on the plate.

油封鸭腿

Oil-simmered Duck Leg

食材 INGREDIENTS

食材	INGREDIENTS	食材	INGREDIENTS
鸭腿 1 个	duck leg 1 PC	橙子皮 10 克	orange 10 g
迷迭香 1 枝	rosemary 1 PC	甜菜头 20 克	beet 20 g
百里香 1 枝	thyme 1 PC	茴香头 20 克	fennel bulb 20 g
大蒜 1 头	garlic 1 PC	红酒酱	red wine sauce
葱 1 根	green onion 1 PC	鸭油适量	duck oil moderate amount
海盐 10 克	sea salt 10 g	沙拉适量	salad moderate amount
胡椒粉少许	pepper powder a little	橄榄油少许	olive oil a little
红酒 10 毫升	red wine 10 ml		

步骤

1. 鸭腿用橙子皮、迷迭香、百里香、大蒜、葱、海盐、胡椒粉、红酒腌制一天一夜。
2. 腌制后放入鸭油中，用50℃左右的温度浸熟。取出后下油锅中将鸭皮煎脆至金黄色。
3. 配上烤好的甜菜头、沙拉，茴香头刨成片，用胡椒粉、海盐、橄榄油拌匀入盘中，配以红酒酱食用。

PROCEDURES

1. Marinate the duck leg with orange peel, rosemary, thyme, garlic, green onion, sea salt, pepper powder, and red wine for 24 hours.
2. Put the marinated duck leg into duck oil, soak in 50 degree oil. Take it out when it is ripe, fry the duck in a pan until the duck skin is crispy and both sides turn golden yellow.
3. Match with roasted beets, salad, and sliced fennel bulbs, add some pepper powder, sea salt and olive oil, and serve with red wine sauce.

小贴士

鸭腿腌好后放进热油里浸泡 5 小时到成熟，油温不能高，不然就成了炸鸭腿。

Tips

Soak the marinated duck leg into hot oil for 5 hours, but the temperature of the oil can not be high, otherwise it will become fried duck leg.

煎鳕鱼 茴香头奶油酱

Fried Cod Served with Fennel Bulb and Cream Sauce

食材	INGREDIENTS
鳕鱼 150 克	cod 150 g
柠檬汁 5 毫升	lemon juice 5 ml
白葡萄酒 5 毫升	white wine 5 ml
海盐 5 克	sea salt 5 g
茴香头 10 克	fennel bulb 10 g
莳萝草 1 根	dill grass 1 PC
奶油 20 克	cream 20 g
胡椒粉少许	pepper powder a little
橄榄油少许	olive oil a little
牛奶适量	milk moderate amount
黑胡椒粉少许	black pepper powder a little

步骤

1. 鳕鱼用黑胡椒、柠檬汁、海盐、白葡萄酒腌制，腌好后放在锅内煎熟。
2. 茴香头切成薄片，加入海盐、胡椒粉、橄榄油拌匀。
3. 另取茴香头切块，放在油锅中炒香，放入牛奶和奶油，煮开取出。
4. 放入搅拌器中打成酱汁，配在煎好的鳕鱼周围，配以莳萝草即可。

PROCEDURES

1. Marinate the cod with black pepper powder, lemon juice, sea salt and white wine, then fry the cod.
2. Cut the fennel bulb into thin slices, add sea salt, pepper powder and olive oil, and mix them well.
3. Cut another fennel bulb into small cubes, fry them in a pan until they are fragrant, add in milk and cream. Take them out after the soup boils.
4. Put them into a blender and smash into sauce, pour them around the fried cod and add some dill grass.

小贴士

煎鳕鱼的时候，注意翻面时不要弄散。

Tips

Don't scatter the cod when the fried cod is turned over.

步骤

1. 三文鱼用海盐、胡椒、柠檬汁、白葡萄酒腌制。
2. 腌制后放在50℃的油温中慢慢浸熟，取出放在空盘中待用。
3. 把芥末酱汁均匀地淋在周围，配以小番茄、可食用的新鲜花草或有辣味的花生碎。

PROCEDURES

1. Marinate the salmon with sea salt, pepper powder, lemon juice and white wine.
2. Soak the salmon in 50 degree oil until ripe, place it on an empty plate.
3. Sprinkle the mustard sauce around the salmon, add small tomatoes, edible fresh flowers or spicy peanut pieces.

小贴士

三文鱼放在油里浸泡，油的温度不能超过50℃，浸泡时间要久些，等慢慢熟透。

Tips

Soak the salmon in oil which is less than 50 degree until it is ripe.

油封三文鱼
芥末酱白酒奶油汁

Oil-simmered Salmon with Mustard Sauce, white Wine and Cream

食材

三文鱼 100 克
番茄 1 个
芥末酱适量
橄榄油 500 毫升
海盐 10 克
胡椒粉 2 克
白葡萄酒 10 毫升
柠檬汁 10 毫升
新鲜花草

INGREDIENTS

salmon 100 g
tomato 1 PC
mustard sauce moderate amount
olive oil 500 ml
sea salt 10 g
pepper powder 2 g
white wine 10 ml
lemon juice 10 ml
fresh plants

奶油水果杯

Cream and Fruit Cup

食材	INGREDIENTS
淡奶油 120 克	light cream 120 g
白糖 10 克	sugar 10 g
吉利丁 4 克	gelatine 4 g
各种水果切小丁	small cubes of all kinds of fruits
白兰地酒适量	brandy moderate amount

步骤

1. 吉利丁泡水后，沥干水分，加白兰地酒溶化。
2. 淡奶油加糖打发后加入溶化的吉利丁，混合均匀，盛进杯子内，放入切好的水果。

PROCEDURES

1. Drain the gelatine after soaking the gelatine in water, melt it with brandy.
2. Put sugar into light cream, whisk them well, add melted gelatine, mix well, fill them in the cup and put in the chopped fruits.

小贴士

吉利丁的比例要适量，水果杯的口感才能最好。

Tips

The proportion of gelatine should be appropriate to make the fruit cup tasty.

焦糖苹果

Caramelized Apple

食材	INGREDIENTS
苹果 1 个	apple 1 PC
白糖 30 克	sugar 30 g
白兰地酒 10 毫升	brandy 10 ml
巧克力饼干碎 30 克	chocolate biscuit pieces 30 g

步骤

1. 苹果去皮，清洗干净，蘸匀白糖，入平底锅内煎成焦糖色，洒白兰地酒。
2. 巧克力饼干碎放盘子内，摆上焦糖苹果。

PROCEDURES

1. Wash and peel the apple, cover it with sugar, fry it in a pan until it turns caramel color, and sprinkle brandy on it.
2. Place the chocolate biscuit pieces on the plate, and place the caramelized apple.

小贴士

注意：糖不要熬煳。

Tips

Don't make the sugar burnt.

焦糖香蕉

Caramelized Banana

食材	INGREDIENTS
香蕉 1 个	banana 1 PC
白糖 30 克	sugar 30 g
白兰地酒 10 毫升	brandy 10 ml
巧克力饼干碎 30 克	chocolate biscuit pieces 30 g

步骤

1. 香蕉去皮，清洗干净，蘸匀白糖，入平底锅内煎成焦糖色，洒入白兰地酒。
2. 巧克力饼干碎放盘子内，摆上焦糖香蕉。

PROCEDURES

1. Peel the banana and wash it, cover it with sugar, fry it in a pan until it turns caramel color, and sprinkle brandy on it.
2. Place the chocolate biscuit pieces on the plate, and place the caramelized banana.

小贴士

可以用这个方法制作焦糖苹果、焦糖橙子等水果。

Tips

You can use this method to make caramelized apple, caramelized orange and other fruit.

煮鳕鱼 南瓜泥焦糖洋葱

Boiled Cod with Pumpkin Puree and Caramelized Onion

食材 INGREDIENTS

食材	INGREDIENTS
鳕鱼适量	cod moderate amount
洋葱	onion
南瓜 20 克	pumpkin puree 20 g
香茅 1 枝	citronella 1 PC
青豆几颗	green bean a few
白糖 10 克	sugar 10 g
黑胡椒 3 克	black pepper 3 g
海盐 5 克	sea salt 5 g
柠檬汁 5 毫升	lemon juice 5 ml
白葡萄酒 4 毫升	white wine 4 ml

步骤

1. 青豆用水煮熟。
2. 南瓜用水煮熟打成南瓜泥。
3. 洋葱做成焦糖洋葱。
4. 鳕鱼用海盐、白糖、黑胡椒、柠檬汁、白葡萄酒腌制，再用保鲜膜包好，入热水中煮熟后放到盘中，配以做好的青豆和南瓜泥、香茅即可。

PROCEDURES

1. Boil the green beans in water.
2. Boil the pumpkin and smash it into puree.
3. Cook caramelized onion.
4. Marinate the cod with sea salt, sugar, black pepper, lemon juice and white wine. Pack the marinated cod with plastic wrap, boil it in hot water until ripe, put it on a plate, and match cooked green beans, pumpkin puree and citronella.

鳕鱼用水煮，一定要包好，避免破烂。

Tips

Wrap the cod well before boiling it in water.

橙子磅蛋糕

Orange Cake

食材	INGREDIENTS
黄油 80 克	butter 80 g
糖粉 80 克	sugar powder 80 g
低筋面粉 100 克	low gluten flour 100 g
鸡蛋 2 个	egg 2 PC
泡打粉 2 克	baking powder 2 g
橙子 1 个	orange 1 PC
时令新鲜水果	fresh seasonal fruits
糖浆部分	syrup
细砂糖 15 克	berry sugar 15 g
水 25 毫升	water 25 ml

步骤

1. 橙子洗干净，皮擦成橙子皮屑（不要擦到白色部分，会苦），加入5克糖。橙子切半，榨汁备用，黄油切小块室温放置后使用。低筋面粉和泡打粉混合后过筛，用刮刀切拌混合。
2. 加入橙汁，混合好面糊，倒入磅蛋糕模具中。
3. 用勺子将蛋糕抹成中间凹两边高。放入烤箱中层180℃烤40~45分钟，至蛋糕顶部焦黄，牙签从裂缝中插进去没有带出湿面糊就可以了，刷蛋糕的糖浆。
4. 糖浆的制作：在平底锅中加入15克糖和25克水，熬煮至沸，立即关火。

PROCEDURES

1. Wash the orange, rub the peel into pieces (don't rub it till the white part, or it tastes bitter), and add in 5 grams of sugar. Cut the orange into halves and make orange juice in a blender. Cut the butter into small cubes for use. Mix the low gluten flour and baking powder, sift them and mix them with a scraper.
2. Add orange juice, mix well, and pour into a cake mould.
3. Use a spoon to rub the cake into a shape that the middle is low and the periphery is high. Put it in the middle of a 180 degree oven for 40 to 45 minutes. Bake the top of the cake brown, when the toothpick is inserted into the cracks, there is no paste on it. Brush the cake with the syrup.
4. Syrup: Put 15 grams of sugar, 25 grams of water into a pan. Boil until the water boils. Turn off the heat immediately.

迷迭香烤苹果

Rosemary-baked Apple

食材	INGREDIENTS
苹果 1 个	apple 1 PC
迷迭香 1 枝	rosemary 1 PC
枫叶糖浆	maple syrup
白糖 20 克	sugar 20 g
树莓 20 克	raspberry 20 g
水 20 毫升	water 20 ml

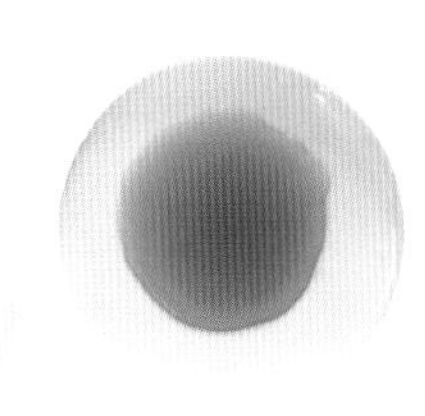

步骤

1. 苹果去皮，裹匀枫叶糖浆，蘸上白糖，撒上迷迭香，入烤箱160℃烤30分钟。
2. 树莓加水和糖熬煮，打成酱汁，搭配烤好的苹果。

PROCEDURES

1. Peel the apple, cover it with maple syrup and sugar, sprinkle some rosemary and bake it in a 160 degree oven for 30 minutes.
2. Boil the raspberry with water and sugar, smash it into sauce to match the baked apple.

小贴士

可以用这个方法烤其他质地较硬的水果。

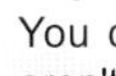

Tips

You can use this method to bake other fruits that aren't soft.

巧克力草莓

Chocolate Strawberry

食材 INGREDIENTS

新鲜草莓 fresh strawberry
巧克力酱 chocolate cream

步骤

将巧克力酱溶化，草莓用竹签串上，蘸匀巧克力酱自然冷却成型，摆进餐具内。

PROCEDURES

Melt the chocolate cream, cluster the strawberry with bamboo sticks and cover them with chocolate cream, place them into tableware after cooling off.

小贴士

溶化巧克力的盆一定不能有水珠，且溶化时只能顺一个方向搅拌。

Tips

The pot used to melt chocolate can not contain water. Stir the chocolate toward one direction when the chocolate is melted.

奶油布丁
Cream Pudding

食材	INGREDIENTS
纯牛奶 700 毫升	pure milk 700 ml
淡奶油 700 克	light cream 700 g
糖 60 克	sugar 60 g
果冻粉 20 克	jelly powder 20 g
香草粉 2 克	vanilla powder 2 g
树莓酱 5 克	raspberry sauce 5 g
巧克力饼干碎 10 克	chocolate biscuit pieces 10 g

步骤

1. 牛奶、淡奶油和糖加热至80℃，加果冻粉打散，确认完全溶解后离火，加入香草粉搅拌均匀，倒入杯子内冷藏备用。
2. 取冷藏成型的布丁，将树莓酱盛进杯子内，用巧克力饼干碎铺在布丁周围。

PROCEDURES

1. Heat the milk, light cream and sugar to 80 degree, add jelly powder and whisk. When the jelly powder melts, turn off the heat, add vanilla power, mix well and pour it into a cup and store it in refrigerator.
2. Take out the refrigerated pudding, put the raspberry sauce into the cup, and put the chocolate biscuit pieces around.

小贴士

也可以在容器底部或中部放入一些水果丁，增添风味。

Tips

Some chopped fruit can be put in the bottom or middle of the container to add some flavor.

步骤

1. 鲈鱼杀好，取肉，切成块状。用海盐、胡椒粉、柠檬汁、白葡萄酒腌制。
2. 腌制后热锅冷油下锅煎制，煎至两面鱼皮金黄变脆取出。
3. 番茄、黑橄榄对切开，锅内加少许油翻炒，炒熟后放入盘中垫底，蔬菜上放煎制好的鲈鱼即可。

PROCEDURES

1. Kill the bass, and cut the fish into cubes. Marinate the bass with sea salt, pepper powder, lemon juice and white wine.
2. Fry the bass in a hot pan with cold oil until both sides turn golden yellow.
3. Cut the tomato and black olive into halves, stir-fry them with a little oil. After cooking, put them on the plate, and place fried bass on the top.

小贴士

煎鲈鱼鱼皮这面时用小火煎熟。

Tips

Use low flame to fry the bass with the skin.

煎鲈鱼番茄黑橄榄

Fried Bass with Tomato and Black Olive

食材

鲈鱼 100 克
番茄 30 克
黑橄榄 20 克
海盐 10 克
白葡萄酒 10 毫升
胡椒粉 4 克
柠檬汁 5 毫升
沙拉油适量
有机花草

INGREDIENTS

bass 100 g
tomato 30 g
black olive 20 g
sea salt 10 g
white wine 10 ml
pepper powder 4 g
lemon juice 5 ml
salad oil moderate amount
organic plants

炸生蚝 · 迷迭香盐

Fried Oyster with Rosemary Salt

食材	INGREDIENTS
生蚝若干	oyster
迷迭香 2 枝	rosemary 2 PCs
天妇罗粉 30 克	tempura flour 30 g
迷迭香盐适量	rosemary salt moderate amount
水适量	water moderate amount
油 500 毫升	oil 500 ml
柠檬汁适量	lemon juice moderate amount
海盐少许	sea salt a little

步骤

1. 天妇罗粉加水调成糊。
2. 生蚝撒少许柠檬汁、海盐，裹上天妇罗粉糊，放入锅中炸好取出。
3. 撒上做好的迷迭香盐，配以迷迭香即可。

PROCEDURES

1. Add water to the tempura flour and make it into paste.
2. Sprinkle the oysters with lemon juice and sea salt, wrap them with the tempura flour, fried them in the pan, and take them out.
3. Sprinkle rosemary sea salt. Match rosemary.

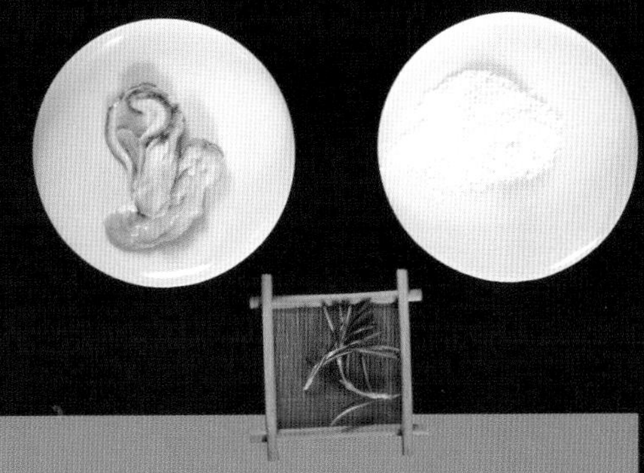

天妇罗粉加水调匀，注意浓稠度，炸好的成品一定是皮脆、肉嫩。

Tips

Mix the tempura flour and water well and pay attention to its thickness. Fried oyster should have crispy skin and tender meat.

图书在版编目（CIP）数据

在家学做法式西餐 ：生活就是一场烹饪 / 余勇浪著；李英涛，那洪伟译. -- 长春 ：吉林科学技术出版社，2018.4

ISBN 978-7-5578-3381-7

Ⅰ. ①在… Ⅱ. ①余… ②李… ③那… Ⅲ. ①西式菜肴－菜谱－法国 Ⅳ. ①TS972.118

中国版本图书馆CIP数据核字(2017)第265467号

ZAI JIA XUE ZUO FASHI XICAN : SHENGHUO JIUSHI YI CHANG PENGREN

在家学做法式西餐：生活就是一场烹饪

著 余勇浪
译 李英涛 那洪伟
出版人 李 梁
责任编辑 端 木 李思言
封面设计 长春市一行平面设计有限公司
制 版 长春美印图文设计有限公司
开 本 710 mm×1000 mm 1/16
字 数 200千字
印 张 12
印 数 1-5 000册
版 次 2018年4月第1版
印 次 2018年4月第1次印刷

出 版 吉林科学技术出版社
发 行 吉林科学技术出版社
地 址 长春市人民大街4646号
邮 编 130021
发行部传真／电话 0431-85635176 85651759 85635177
85651628 85652585
储运部电话 0431-86059116
编辑部电话 0431-85635186
网 址 www.jlsycbs.net
印 刷 长春新华印刷集团有限公司

书 号 ISBN 978-7-5578-3381-7
定 价 59.90元